Local Science vs. Global Science
Approaches to Indigenous Knowledge in International Development

Edited by

Paul Sillitoe

Berghahn Books
New York • Oxford

First published in 2007 by
Berghahn Books
www.berghahnbooks.com

© 2007, 2009 Paul Sillitoe
First paperback edition published in 2009

Library of Congress Cataloging-in-Publication Data

Local science vs. global science : approaches to indigenous knowledge in
international development / edited by Paul Sillitoe.
   p. cm. -- (Studies in environmental anthropology and ethnobiology ; v. 4)
Includes bibliographical reference and index.
ISBN 1-84545-014-0 (hardback : alk. paper)
   1. Ethnoscience--Developing countries. 2. Indigenous peoples--Ecology--
Developing countries. 3. Technical assistance--Anthropological aspects--
Developing countries. 4. Community development--Development countries.
5. Applied anthropology--Developing countries. 6. Natural resources
management areas--Developing countries. I. Sillitoe, Paul, 1949-

GN476.L63 2006
001.089--dc22
                                                                2006019694

British Library Cataloguing in Publication Data

A catalogue record for this book is available from the British Library

Printed in the United States on acid-free paper

ISBN 978-1-84545-014-4 (hardback)
ISBN 978-1-84545-648-1 (paperback)

Local Science vs. Global Science

Studies in Environmental Anthropology and Ethnobiology

General Editor: **Roy Ellen**, FBA
Professor of Anthropology, University of Kent at Canterbury

Interest in environmental anthropology has grown steadily in recent years, reflecting national and international concern about the environment and developing research priorities. This major new international series, which continues a series first published by Harwood and Routledge, is a vehicle for publishing up-to-date monographs and edited works on particular issues, themes, places or peoples which focus on the interrelationship between society, culture and environment. Relevant areas include human ecology, the perception and representation of the environment, ethno-ecological knowledge, the human dimension of biodiversity conservation and the ethnography of environmental problems. While the underlying ethos of the series will be anthropological, the approach is interdisciplinary.

Volume 1
The Logic of Environmentalism:
Anthropology, Ecology and Postcoloniality
Vassos Argyrou

Volume 2
Conversations on the Beach:
Local Knowledge and Environmental Change in South India
Götz Hoeppe

Volume 3
Green Encounters:
Shaping to Indigenous Knowledge in International Development
Luis A. Vivanco

Volume 4
Local Science vs. Global Science:
Approaches to Indigenous Knowledge in International Development
Edited by Paul Sillitoe

Volume 5
Sustainability and Communities of Place
Carl A. Maida

Contents

List of Figures

List of Tables

Acknowledgements

This book originates from the British Association's Festival of Science at Salford in 2003 where the 'Anthropology and Archaeology Section' and 'General Section' arranged coordinated sessions entitled 'Local Science versus Global Science' and 'Traditional Knowledge and Modern Science'. I am grateful to David Shankland, then Recorder to the 'Anthropology and Archaeology Section', for all of his support and encouragement, without which the event would not have been anything like as successful. I also thank David Dickson, the Recorder to the 'General Section', for coordinating his session with us and kindly agreeing to some of the contributors publishing their papers in this volume. Other contributions come from papers presented at the Association of Social Anthropologists' Decennial Conference at Manchester in 2003 in a panel entitled 'Beyond Science: Approaches to Local Knowledge in Development'. I thank all those who participated in these events for their intellectual support. I also thank my wife Jackie for her unfailing assistance and encouragement. For generous financial support of the events at which colleagues presented the papers in this volume, we thank the British Academy, British Association, Royal Anthropological Institute, Association of Social Anthropologists of the U.K. and Commonwealth, Arts and Humanities Research Board, and the Wenner-Gren Foundation.

List of Contributors

Alberto Arce
Senior Lecturer, Department of Sociology of Rural Development, Wageningen University, Hollanseweg 1, 6706 KN Wageningen, The Netherlands.

Gerard Bodeker
Division of Medical Sciences, University of Oxford, Oxford OX2 6HG, U.K., and Adjunct Professor of Epidemiology, Columbia University, U.S.A.

Marina T. Campos
Doctoral Candidate, School of Forestry and Environmental Studies, Yale University, Sage Hall, 205 Prospect Street, New Haven, Connecticut 06511, U.S.A.

David A. Cleveland
Environmental Studies Program and Center for People, Food and Environment, University of California, Santa Barbara, CA 93106-4160, U.S.A.

Charles Clift
Secretary, Commission on Intellectual Property Rights, Innovation and Public Health (CIPIH), Room E127, World Health Organization, 21 Avenue Appia, 1211 Geneva 27, Switzerland.

Michael R. Dove
Margaret K. Musser Professor of Social Ecology, School of Forestry and Environmental Studies, Yale University, Sage Hall, 205 Prospect Street, New Haven, Connecticut 06511, U.S.A.

Roy Ellen
Professor of Anthropology and Human Ecology, Department of Anthropology, Marlowe Building, University of Kent at Canterbury, Canterbury, Kent CT2 7NR, U.K.

Eleanor Fisher
Lecturer in International Development, Centre for Development Studies, University of Swansea, Singleton Park, Swansea SA2 8PP, Wales, U.K.

Serena Heckler
Research Associate and part-time lecturer, Department of Anthropology, University of Durham, 43 Old Elvet, Durham DH1 3HN, U.K.

Geoff Howell
Formerly Environment Canada, Atlantic Canada Region, Canada.

Mariella Marzano

Research Associate, Department of Anthropology, University of Durham, 43 Old Elvet, Durham DH1 3HN, U.K.

Andrew S. Mathews
Assistant Professor of Anthropology, Department of Sociology and Anthropology, Florida International University, DM336A, University Park, Miami FL 33199, U.S.A.

Virginia Nazarea
Professor of Anthropology, Department of Anthropology, University of Georgia, Athens, GA 30602-1619, U.S.A.

Peter Penashue
Past President of the Innu Nation of Labrador, P.O. Box 169, Northwest River, Newfoundland/Labrador A0P 1M0, Canada.

Anne Rademacher
Assistant Professor and Faculty Fellow, Department fo Social and Cultural Analysis, New York University, 41 East 11th Street, RM723 New York, NY10003, U.S.A.

Steve Rhee
Doctoral Candidate, School of Forestry and Environmental Studies, Yale University, Sage Hall, 205 Prospect Street, New Haven, Connecticut 06511, U.S.A.

Robert E. Rhoades
Distinguished Research Professor of Anthropology, Department of Anthropology, University of Georgia, Athens, GA 30602-1619, U.S.A.

Trudy Sable
Director, Labrador Project, Gorsebrook Research Institute Saint Mary's University, Halifax, Nova Scotia, Canada.

Paul Sillitoe
Professor of Anthropology, Department of Anthropology, University of Durham, 43 Old Elvet, Durham DH1 3HN, U.K.

Benjamin R. Smith
Research Fellow, Centre for Aboriginal Economic Policy Research, Hanna Neumann Building, The Australian National University, Canberra, ACT 0200 Australia.

Daniel S. Smith
Assistant Professor of Environmental Studies, Ramapo College, 505 Ramapo Valley Road, Mahwah, New Jersey 07430, U.S.A.

Daniela Soleri
Research Scientist, Environmental Studies Program and Geography Department, University of California, Santa Barbara, CA 93106-4160, and Co-Director, Center for People, Food and Environment, Santa Barbara, CA 93110, U.S.A.

Dave Wilson
Formerly Project Manager, Environment Canada, Atlantic Canada Region; currently 55 Purcell's Cove Road, Halifax, Nova Scotia B3P 2G3, Canada.

Laura M. Yoder
Syiah Kuala University, P.O. Box 567, Bareda Aceh, NAD 23001, Indonesia.

1 Local Science vs. Global Science: an Overview

Paul Sillitoe

Relativity is a relative idea. There is the physical scientists' notion of relativity, which is of global relevance, and there is the social scientists' notion of relativity, which is of local relevance. Just as physicists argue that classical scientific laws are a special case that apply to planet Earth only – varying with mass, speed, time and such – so too anthropologists maintain that scientific theories are a special case rooted largely in contemporary Euro-American understanding of the world – varying with culture, history, place and so on. We assume that our scientific view is one way of explaining our experience of the world, albeit a technically powerful one, employing an astonishingly effective body of integrated theory. Few of us are arrogant or ignorant enough to think that humans can aspire to a true 'Godlike view' of the universe and our place in it. We can never definitively verify universal propositions, only unsuccessfully attempt to falsify them experimentally (Popper 1959).

Increasingly there are calls that we should pay more attention to other views, not necessarily to challenge scientific achievements but to inform them further and better guide their exploitation, to enrich their cultural relevance to the benefit of all. The promotion of alternative views is burgeoning in the context of international development currently. The aim is to find a place for others' knowledge alongside science (Richards 1985; DeWalt 1994; Antweiler 1998; Purcell 1998; Sillitoe 1998; Ellen and Harris 2000; Sillitoe et al. 2002; ISSJ 2002). The distinctive feature of these essays is that they treat local and global science as culturally equal, tackling head on the import of purported differences, acknowledging both strengths and weaknesses, and opportunities for further synergistic interaction. Advocates of local knowledge argue that we need to learn to listen to others. One reason for the endorsement of so-called local or indigenous knowledge initiatives in the context of participatory development is that these will likely facilitate more successful interventions. They promote culturally appropriate initiatives and empower people to contribute to their formulation and implementation.

In anthropology there is a long tradition of advocating the soundness of local science. For instance, over a century ago the Master of Hatfield College in Durham, in an address to the University's Philosophical Society entitled 'Savage Science', observed that 'the foundation, the principle, and the methods of savage logic and scientific logic are identical … modern science has no other basis than savage science had – both are built on the same foundations and by means of the same instruments of thought' (Jevons 1900: 12, 19). We now have many examples of the soundness of local science and practices, and the need to respect them – some previously considered 'primitive' and in need of change. We have long known that shifting cultivation, far from destroying tropical forests, is a sustainable land use regime (Conklin 1957), and with appropriate crop and soil management it can transform into stable semi-permanent farming systems, as in highland New Guinea (Sillitoe 1996; see also Dove et al., this volume). Shifting cultivators in the Himalayan foothills are not the reckless farmers that some assert, responsible for land degradation and massive soil erosion losses leading to lowland sedimentation and water shortages, but are aware of the environmental risks and farm in a regulated way to conserve their soil resources (Forsyth 1996). The close investigation of West African forest-savannah land use history reveals that where previously people were accused of creating savannah from forest they are found to operate sophisticated woodland management regimes (Fairhead and Leach 1996). The trampling of vegetation and soil by cattle in East Africa, once cited as evidence of pastoralist over-stocking degrading rangeland, has been revealed as sound management promoting the vigorous growth of new pasture (A. Kassam, pers. comm.; cf. Leach and Mearns 1996). Similarly, Aboriginal bush-burning practices, long condemned as environmentally destructive, are now recognised as sound management adopted in Australian national parks for clearing overgrown and dead vegetation and ensuring healthy new growth (Verran 2002; see also Smith, this volume). The introduction of high-yielding rice varieties into Bali, with associated chemical inputs and changes to the farming regime, had deleterious consequences because technocrats did not appreciate how the sophisticated irrigation system operated, centring on networks of water temples and rituals that synchronised cultivation across the island (Lansing 1991). The crop-breeding activities of peasant farmers, far from being hit or miss as popularly imagined, match well those of plant geneticists, and have proved highly effective for millennia to judge by the number of cultivars farmers have domesticated, selecting for particular environments (Richards 1986; Cleveland and Soleri 2002) Attempts to improve on the 'primitive' fishing technology used on the African Great Lakes likewise proved detrimental to fish stocks, the local 'inefficient' practices being well adjusted to the conservation of fish populations (E. Allison, pers. comm.). There are many other examples of the soundness of local science.[1]

We resist the urge to define global and local science, beyond making a few general observations, for reasons that are discussed in more detail below, namely the difficulty of coming up with a definition that covers the multitude of local sciences sufficient to contrast with global science. It is established, as Jevons's comments intimate, that all humans are capable of abstract thought and have notions of causality, that they can suspend prior beliefs and will revise these if evidence suggests that they are wrong, even if counter-intuitive. We broadly hold to the dictionary (*Oxford English Dictionary*) definition of science as an intellectual and practical pursuit that seeks to further understanding of the 'physical and natural world through observation and experiment'. In extending this definition to local contexts we have to interpret behaviour broadly, such that experiment includes noting the results of everyday experiences – for example, while subsistence farmers may not consciously hold a formal hypothesis and conduct randomised trials to test it, they continually experiment as they cultivate crops and learn from the results, passing on the knowledge in accumulated lore. While local scientific knowledge may not be systematically recorded like that of global science, nor feature such prescribed theories, it is nonetheless formalised to varying extents in cultural heritage. All cultures accumulate and interpret knowledge rationally according to their value codes, although until we appreciate these latter it may seem otherwise. We believe that local and global science are often directly comparable. The implication is that local practices may validate narrowly defined global or 'modern scientific' knowledge, as in testing and adopting or rejecting extension advice for example (see Marzano, this volume), and vice versa, that global scientific techniques may corroborate local knowledge, as in for instance some ethnoscientific research (Sillitoe 1996). While some contributions to this volume are concerned to compare and contrast local with global science, and repeatedly broach these definitional issues, others are more interested in the relations between local and global science, particularly their political dimensions. According to Ingold's (2004: 177) tripartite definition of anthropology's engagement with science, the essays fall into studies of science – as a way of working and kind of knowledge – and the relation between science and society: the impact of science on people and their responses to it.

The learning process should be a two-way affair, not only facilitating the adoption of scientifically informed ideas by local communities but also the informing of scientific understanding with local knowledge. There is nothing new in this proposition. Eighty years ago Malinowski, reputedly inaugurating the modern era in anthropology, argued on the last page of *Argonauts of the Western Pacific* (1922: 518), that 'In grasping the essential outlook of others ... we cannot but help widening our own ... (it) should lead us to such knowledge and to tolerance and generosity, based on the understanding of other men's point of view.' The contribution of Gerard Bodeker illustrates this, showing how the randomised control trials that characterise clinical medicine's hunt for new drugs diverge from local practices and may result in modern medicine overlooking valuable local knowledge. He describes how Indian (Ayurvedic) and Chinese medical practitioners consider the whole person in disease diagnosis and not only a few symptoms; they give priority to state of mind and talk about balancing energy forces. Likewise their pharmacologies work on the synergistic interaction between several plant components in a remedy, based on a profound knowledge of herbal plant taxonomy, which is entirely different to modern science that seeks to isolate the active chemical compound in any one plant and subsequently synthesise it. It is an example of the holistic approach of indigenous knowledge systems in contrast to the reductionistic approach of science. He argues that modern medicine should take account of such complex combinations of plant materials, using the case of *changshan*, an ancient Chinese antimalarial drug, to make his point. The active ingredient in *changshan* is a saxifrage (*Dichroa febrifuga*), reported to be one of two of the most powerful antimalarial compounds known. Western science identified and extracted the active alkaloid but rejected it as it caused severe nausea. Herbalists administer it in a formulation with six other ingredients. If clinical medicine had considered the traditional herbal mix, which includes ingredients to offset nausea, it could have developed an effective drug to combat malaria years ago. This collection of essays intends to go beyond such demonstrations of the soundness of local science, and arguing for the incorporation of such knowledge into development alongside techno-scientific work to reduce poverty, to contend that we need to look quizzically at the foundations of global science itself and further challenge its hegemony, not only over local communities in Africa, Asia, the Pacific or wherever, but also the global community.

Local Knowledge Informing Science?

The reason for global science's success is that it has achieved what appears to be a good approximation to understanding natural processes, notably in terms of the ability that it affords humans to manipulate and control the physical world, as evidenced in the startling technological achievements that it underpins. We have only to think of air and space travel, the electronic wizardry of computer and robotic technology, organ transplant surgery, the increasing ability to intervene in life itself with test-tube conception, genetically modified crops, and so on. Such scientific advances are regularly announced and celebrated in events such as the annual British Association Festival of Science meetings from which this volume originates. It is such technological achievements, undoubtedly the most advanced in human history, that have allowed Euro-American society its global domination, from trading colonialism to multinational corporatism, and have led to the muting of other cultural views and values, even threatening their continued existence. Some may assert that this verifies the assumptions of evolution, one of science's most popular theories, for we see here the survival of the fittest cultures. The problem is defining cultural fitness, for evolution is not a teleological theory. In the long term, if we are not careful, our scientific-technological domination may prove unfit or unsustainable, as some of the contributions to this volume point out (see chapters by Cleveland and Soleri, Rhoades and Nazarea, and Sillitoe).

The idea that others' knowledge might have something to contribute – even challenge scientific understanding – appears to many a hopeless, even a silly proposition. It seems to be a David and Goliath scenario: no other culture has come close to science's aeronautical, communications, medical and other technological achievements. Yet experience shows, as the few examples given above illustrate, that local people often get it right, sometimes when science gets it wrong. Local views, with sympathetic research, can enrich scientific understanding. This is not a revolutionary observation. The history of science testifies to it. Folk knowledge from Europe and elsewhere has informed the development of science. In their contribution Alberto Arce and Eleanor Fisher point out how everyday understanding and practice has influenced science since its Enlightenment emergence – when 'natural knowledge' was used to encompass 'skill or craft' and subsequently 'the workings of nature' – to the present day – when sociologists argue that science is not culturally disembodied knowledge but prejudiced by social factors such as worldview, verbal categories, semantics and shared practices. The mathematics that underpins the quantitative foundations of much natural science (see Chapter 13, this volume) owes much to medieval Islamic scholarship (Nasr 1976). Since European global expansion, we have drawn on the knowledge of local populations to further understanding of new plants and animals (Atran 1990; Turnbull 2000): as seen, for example, in Rumphius' flora of southeast Asia, which drew heavily on native classifications and ecology that, picked up by Linnaeus, continue to influence scientific biology to this day (Ellen and Harris 2000: 6–11; Ellen 2004). Local knowledge contributed to the emergence of the theory of evolution, one of science's most successful theories, Galapagos inhabitants pointing out to Darwin (1845) that they could tell from which island tortoises came, and finches too, by their markings and shape, the result of micro-evolution. And currently in this vein, we have pharmaceutical companies actively prospecting others' knowledge in their search for new drugs and cosmetics, continuing a tradition that has already resulted in over 25 percent of Western prescription drugs deriving from indigenous knowledge (Puri 2001).

This history makes puzzling the scientific establishment's apparent scepticism regarding growing interest beyond anthropology in indigenous knowledge, notably in international development. An editorial in the influential journal *Nature* (October 1999), captioned 'Caution: Traditional Knowledge. Principles of Merit Need to Be Spelt out in Distinguishing Valuable Knowledge from Myth', catches the tone. It was commenting on the 1999 World Conference on Science, organised in Hungary by the International Council for Science and UNESCO (United Nations Educational, Scientific and Cultural Organization), on 'Science for the Twenty-First Century: A New Commitment', at which representatives of the U.K.'s Royal Society and the U.S. National Academy of Sciences, and subsequently the International Council's general assembly, questioned some of the clauses on local knowledge in the *Declaration on Science* and *The Science Agenda* adopted by the conference. A controversial clause in the former refers to 'traditional and local knowledge systems as dynamic expressions of perceiving and understanding the world, [that] can make, and historically have made, a valuable contribution to science and technology, and that there is a need to preserve, protect, research and promote this cultural heritage and empirical knowledge' and the latter calls on governments to 'formulate national policies that allow a wider use of the applications of traditional forms of learning and knowledge' (Dickson 1999: 631; Martin 2002; http://helix.nature.com/wcs).

One reason for concern on the part of the scientific establishment was that it perceived support for pseudo-science, such as beliefs in creationism, which could impact negatively on it, as Christian fundamentalism recently has in parts of the U.S. (Deveraux and Evans 2004). Another, some argue, is the concern that scientists have to protect their status and power in society, and the global hegemony of Western culture, by maintaining a clear boundary between what they know and do, from what everyone else knows and does (Nader 1996). Although the cultural location of the hegemony is increasingly unclear (historically science as narrowly defined emerged in Europe between 1600 and 1900), it is now a global phenomenon involving many scientists from non-Western backgrounds. Yet another reason is irritation on the part of some scientists with postmodern sociological commentaries that point out that their work is culturally relative and subjective, not the objective and value-free quest for truth that they think (Knorr-Cetina 1999). And related to this, there is some annoyance at attempts by historians of science to present the human side of scientific endeavours – the accidents, nonsense, blind alleys etc. – that many scientific texts omit when they present theories as abstract deductions, depicting these in abstract equations such as $e = mc^2$ and omitting the vivid metaphors about trains travelling at the speed of light (for example see Spranzi's 2004 entertaining account of Galileo's use of metaphors). Others resent sociologists prying into their lives to determine the significance of social interaction and values in determining scientific outcomes and their application in technology– the role of gossip, 'old school tie', personal likes and animosities within and between laboratories etc. (Latour and Woolgar 1986; Latour 1987, 1999). Some contributions to this volume take up these social and political issues (see Arce and Fisher, Dove et al., Marzano, and Sable et al.), albeit with the intention of contributing to the rapprochement of local with global science, not exacerbating these unhelpful relations.

Local and Global Science Compared

The contrasting of local with global science is not necessarily straightforward, as several of the chapters in this book show, and intimated in Bodeker's discussion of the epistemological clash between 'traditional' and modern medicine. We need to approach local schemes with sympathy if we are to learn from them. In their contribution Bob Rhoades and Virginia Nazarea compare scientific modelling of environmental change with local perceptions of change in the Cotachachi-Cayapas Ecological Reserve in Ecuador.[2] Land use management in the region employs a computer model to predict future changes from current trends to inform policy makers and politicians in making development decisions. The modellers assume that time is linear and that cause-and-effect relations mark its passage, which contrasts with the Quechua people's conception of the future being behind them while the past is in front, such that dealing with the unseen future entails seeing the past clearly. In my contribution I also pick up on science's linear perspective, contrasting it with circular ideas elsewhere. It is difficult to see what we have to learn from places such as New Guinea, where a wide range of counting systems occur, all featuring a limited number line and restricted series of terms for numerals (in some cases only two words), and which feature repetitive counting in rounds, starting again at the beginning when they reach the end of their limited number sequence. They make calculation seem even more unwieldy by sometimes reckoning in pairs and employing numeral classifiers, which are words attached to numbers depending on the thing counted, such as calling circular things 'round', and when counting coins saying 'round one, round two, round three' etc. We are inclined to dismiss such numerical schemes when comparing them to the sophisticated mathematical logic and computational power of the linear decimal scheme used in global science, when their ontological assumptions reflect radically different worldviews and insights that demand more respect.

Serena Heckler illustrates the shortcomings of scientifically framed research when confronted with ethnographic reality. The moral for a volume such as this is beware of scientific assumptions distorting local understandings. Her hypothesis was that Piaroa communities more acculturated into mainstream Venezuelan society would have less botanical knowledge than remote forest ones. She set up a number of quadrat plots to test it, where she asked people to identify plants and give their uses, collecting quantitative data for analysis by scoring respondents' replies according to correct responses. Subsequently reflecting on the research, she came to question the results when she realised that Piaroa botanical nomenclature is mutable, people even inventing plant names for her. She frankly reports how she came to realise that variation between individuals and fluidity of classification can characterise such people's knowledge of the forest and compromise scientific assumptions. However not all local ideas are necessarily at odds with scientific understanding; they may parallel it. The contribution of Trudy Sable and colleagues shows how local and global science can profitably inform each other. They report on a collaborative project[3] that involves both natural and social scientists working together with Innu people in Labrador to research locations where frozen lakes first thaw in the spring to reveal water rich in fish and attracting much wildlife, especially waterfowl. They show how the results of water sample analysis and limnological research undertaken by scientists compare with Innu ideas of water quality and health, and their understanding of the ecology of waterbodies and how they exploit these resources.

A related issue is the local character of much local knowledge. This is proving a problem in development contexts where agencies seek generic solutions to problems of poverty, sustainability and so on. We see the scale problem, as Cleveland and Soleri point out, in the increasing focus of natural resources development on diverse regions with marginal and fragile environments away from those where high production is achievable using a standard technology, usually including high-yielding varieties. By definition local science is geographically and culturally specific, infrequently extending to wide regions. We see this in Australia, as Ben Smith makes clear, comparing Cape York Aboriginal plant knowledge with botanical science. Knowledge for these Aboriginals is so localised that only persons with kin and spiritually validated associations with a place will profess to know about it. They do not allow for general categories divorced from place; for instance they will not classify together two spears made in different places because they have dissimilar 'immanent essences' resulting from different creative acts of 'culture-heroes' in the mythical Story- or Dreamtime. In some senses local and global scientific ideas are incommensurable. For example, an Aboriginal elder and a geological scientist looking at a rock formation will 'see' quite different things: the Aboriginal elder sees the petrified record of some event in the Story-time involving some creator being such as the rainbow serpent, whereas the geological scientist sees a record of sedimentary processes millennia ago and contemporary weathering activity (and if employed by a mining company possible evidence of certain minerals in the outcrop). It is not a case of one being right and the other wrong – as some scientists might contend – but of radically different worldviews and epistemologies.

The contribution by Roy Ellen illustrates well how the regional focus of local knowledge contrasts with the universal sweep of global science in a comparison of an Indonesian population's classification and understanding of forest types with that of foresters and ecologists. The Nuaulu of Seram island have a locally focused scheme, as evidenced in the extensive use of toponyms that convey information about places' histories and how human activities have modified their vegetation. An investigation of the composition of a series of forest quadrat plots, relying heavily on local knowledge of forest communities, shows that they take several factors into consideration in classifying forest and do so in a flexible way, acknowledging its continuous variation. This contrasts with global science that wishes to generalise for the world, focusing not on the forest cover of a small island, but all forest throughout Maleasia or even the tropics; although comparing categories used in one region of the world – e.g. equatorial Africa – with those in another – e.g. Amazonia – can be difficult. Until recently this has involved the use of gross forest type categories to accommodate the wide variety of communities found, although sometimes it has drawn on local knowledge, as evident in the scientific mapping of Seram, which features many Nuaulu terms – another example of local knowledge contributing to scientific knowledge over a long period of time. The equilibrium model of static primary or climax forest, on which depend such stereotype classifications of forest according to ideal types is giving way, no longer accepted by ecologists who seek to understand the detailed structural diversity of forests. Again, it is not necessarily a case of incommensurability, as new computing technology is increasingly allowing global science to work with a multitude of different forest communities, bringing it nearer to the toponym knowledge of local populations and its more faithful representation of forest variation.

Hybrid Science

Correspondence between local and global sciences is further evident in the hybrid knowledge that results when they borrow from one another, as has occurred for generations with interaction between populations – what anthropologists once called diffusion. This on-going process even questions the propriety of distinguishing local from global science, prompting a furious debate over the meaning of such terms as local, indigenous, traditional and citizen knowledge, among others, and the correctness of using such terms at all. It smacks for some of the discredited distinction between primitive and civilised thought, what some have dubbed 'the great cognitive divide' (Frake 1983; Ellen 2004: 411), a reflection of our urge to impose simple, and often sterile (as here) dichotomies on the world. In this context it is necessary to point out that while there is a single global science (which is not to imply that scientists always agree with one another), there are a large number of local sciences, illustrating the richness of human inventiveness; to suppose that they reflect different cognitive processes is fallacious, although they do reveal varying preoccupations in life and differing bodies of knowledge. As Roy Ellen (2004: 443) observes, 'what counts as "indigenous knowledge" is so protean and extensive, to claim that it is comparable or not comparable to science is misleading, a diversion from the real issue'. The real issue is dealing with the diversity and dynamism that characterises human understanding. Furthermore, as mentioned previously, and as the label 'global' indicates, many scientists today come from non-Western cultures, enriching narrowly defined science's enquiries. There is a need to engage with the complexity of relations that characterise all knowledge traditions, to break with these fruitless definitional debates, such as I have attempted elsewhere by envisaging linked spheres of knowledge instead of two poles, local and global (Sillitoe 2002).

Several contributions to this volume deal with issues relating to such hybridisation knowledge. The project on which Trudy Sable and colleagues report illustrates well such interaction between local and global science, with the foundation of the Innu Environmental Guardians Program, a university-accredited course driven by Innu needs and values, which seeks to redress the damage done by the European education system and empower the Innu to deal with state and federal government agencies. Innu elders determine part of the programme's contents and act as instructors alongside lecturers in Western environmental science, to achieve the balance the Innu need to manage in twenty-first-century Canada.

Mariella Marzano focuses on the relation between local and global science as evidenced in the role of extension in development – extension comprises the advisory services that inform and teach people about interventions. It illustrates one process of hybridisation of scientific with local knowledge. She argues that we need better to understand why farmers opt for some scientific solutions and not others, which often relates to social and political issues as much as technical ones. She uses experiences in eastern Sri Lanka, working with a project researching intercropping with rubber, to illustrate the problems extension workers face interfacing between scientific advice and local farmer 'realities'. Among the difficulties they face in promulgating scientifically researched solutions is the unsuitability of some of the advice they are expected to disseminate to local farmers. This often relates to ignorance of local knowledge and conditions, because many agricultural research stations continue to work largely in isolation from farmers, subsequently seeking to impose their scientific solutions on them via extension workers.

Michael Dove, Daniel Smith and colleagues offer a salutary warning of the dangers of perpetuating the image of a clear divide between local knowledge and global science, a risk courted by books such as this one. Through a series of ethnographic vignettes they criticise the separation of Western from non-Western resource use systems and environmental relations, and the 'mobilisation' of such environmental ideas to political ends. They remind us of the hybrid nature of much environmental knowledge, that historically mixing has occurred, with the incorporation of global scientific knowledge locally, through a series of case histories. These include the influence of commercial logging companies on community forestry management and fire fighting in Mexico; the impact of mapping on communities in Irian Jaya and the consequences of their participation in the process; and Dayak people in East Kalimantan promoting themselves via an environmentally aware co-operative to attract outside agencies to work with them.

Alberto Arce and Eleanor Fisher explore what they call the interface between knowledge traditions, particularly how everyday understanding and practice influence scientific research in agriculture, focusing on processes where global science confronts and influences, and is influenced by, different local practices. They illustrate the negotiation of knowledge that occurs between different actors at the interface via case histories. These include an account of the experimental trials by Lawes, a farmer, and Gilbert, a chemist, at Rothampstead, which led to inorganic phosphate fertilisers; and the work of Pearl in the U.S. on hen productivity, which increased egg laying by combining a close observation of farmers' practices with a knowledge of genetics, and imaginative use of extension services to relay the information back to farmers. In his contribution Ben Smith describes communication between local Aboriginal and scientific traditions in Australia, regardless of the epistemological gulf between them. The Australian State has coerced Aborigines into European-style education for generations, in misguided attempts at forced assimilation, such that they are able to broker between both traditions. We see this in development projects such as that drawing on indigenous plant knowledge or 'law' to distil commercial oils, to establish an environmentally and socially sustainable local enterprise.

Variable Tacit Knowledge

The fallacy of bracketing together all local science in contradistinction to global science is not only evident when we look at any local tradition closely, as in many of the chapters in this book, but is doubly so when we examine the assumption that the knowledge represented is locally homogenous. The variability in local knowledge is something that we have to accommodate – see my linked spheres of knowledge model (Sillitoe 2002). Different interest groups within a community might have different understandings of issues, with different perspectives and agendas, which they will seek to manipulate, those in more powerful positions usually doing so more successfully, imposing their views on others. Differences will exist along gender, age, class, occupational and other lines, and between individuals of similar social status, although we should beware of overstating the extent to which knowledge varies between people who share a common sociocultural and linguistic heritage. The interpretation they put on shared knowledge will differ, depending on how it affects their interests. It is common for in-fighting to occur between different interest groups within a community regarding proposed development interventions. We see such differences in the global science community too, where scientists regularly disagree over the interpretation of research results (see Cleveland 2001 for an example of how context influences scientific interpretation). Alberto Arce and Eleanor Fisher afford an illustration in their discussion of an agricultural research station – at El Chapare in Bolivia – which, with the building of a new high-tech in vitro crop propagation unit with international aid, not only sought to foist an inappropriate technology on local farmers (with the aim of deflecting them from coca drug cultivation by promoting commercial palmito production), but also spawned political in-fighting among staff and, paying little attention to the realities of farming in the region, predictably failed to help the peasant population.

The challenges do not end here. The local focus of local science reflects not only an absence of any universal aspirations but also the pragmatic rootedness of much of this knowledge, which is contingent on acquiring particular skills necessary to life in certain regions. The documentation of such tacit knowledge presents intriguing problems. People do it, they do not debate it. The only sure way to access such experiential knowledge is to learn to do the activities that comprise it (hence the importance anthropologists accord to participant observation, although we may question the extent and ethics of their participation). Cleveland and Soleri touch on issues of tacit knowledge in their discussion of what they call the 'natural farmer', as do Arce and Fisher in their case histories of agricultural innovations. In her chapter, Serena Heckler reports on the importance of such knowledge, that the Piaroa of Venezuela learn much through their life experiences 'going to' the forest, in what they do and see while engaged in activities there. The importance of shamanistic practice to what these Amazonian Indians know adds another experiential dimension, which further undermined her initial scientifically framed research project. While individuals may be knowledgeable about forest botany from a practical perspective, they may be ignorant of what really matters from a Piaroa perspective, namely the work of gods and spirit forces that ensure forest fertility.

The problems are compounded when we consider the limited and distorted understanding we achieve – for instance of shamanistic forest knowledge – when we try to convey whatever we do manage to learn using the written word. This does not invalidate attempts, however partial, to record something about these matters, but it puts them in perspective. When we engage with tacit knowledge we have to admit that there are dimensions to understanding and living in the world other than the intellectual, for human experience and knowledge encompass far more than words can convey. Academics perhaps over-intellectualise and assume that they can capture too much of the human condition in rational discourse. This takes a particular turn in relation to domains that depend equally on experience as intellect, on skills transferred 'hands-on', which practitioners often cannot verbalise. We sometimes find ourselves in the awkward position of appearing to explain what they themselves cannot explain, while it is their practices and behaviour that embody the knowledge! This is not only an issue with local science. Experimental science also features an element of tacit knowledge: witness how often scientists refer to the importance of their practical skills and those of their laboratory technicians. Many early scientific achievements featured the work of handy persons such as Faraday and Davy building and operating strange gadgets to investigate the properties of electricity (Gooding and James 1985; Fisher 2001). This brings us back to the social and cultural issues touched upon earlier, that any science, local or global, depends on interaction between socially and culturally placed humans who embody differing skills.

Local and Global Science in Socio-cultural Context

Several contributors to this book focus on social and cultural issues, largely their political dimensions, which have been of interest for some time in sociology and history of science debates. Recurring themes include power relations in science and imperialism in development contexts. A worrying aspect of global science for many is the idea that it researches and discovers the 'secrets of nature' in a disinterested manner, that it seeks knowledge and advances understanding for their own sake (Chalmers 1999). Nature is the way she is, according to this view, and all scientists are doing is revealing her more fully to us, and in advancing our understanding are allowing us to exert further control over the world and our destiny (Whewell 1989). Regardless of the discoverers, whatever they uncover will be the same because nature is a constant beyond those who investigate her, such that they will all reveal the same mysteries. But whether or not you believe nature exists out there and awaits uncovering, the discoveries undeniably take place and are interpreted within a certain sociocultural and historical context (Latour 1987, 1999), which has largely been Euro-American capitalist society during the last two centuries that global science has emerged. Scientific research clearly does not take place in some sociopolitically neutral environment, the place where those who quest after 'pure knowledge' live. Place, culture and time heavily inform it, particularly interpretation of scientific findings and the uses to which we put them.

The cases documented by Michael Dove, Daniel Smith and colleagues highlight the use of environmental discourse to manipulate various political agendas. On the one hand, there are authorities deprecating local knowledge and practices to exert control over communities. We see this in the questionable depiction of swidden agriculture as backward and damaging to the environment when this farming regime occurred in parts of Europe until recently and still does in the U.S., which 'anti-swidden forces' conveniently overlook as they seek to dispossess farmers. And the presentation of the Northern Forest in New England U.S. as pristine wilderness by environmental and conservation organisations, when commercial logging and, increasingly, tourism have severely affected it, signify control by powerful outside interests. On the other hand, people are valorising non-Western environmental knowledge and stewardship in the face of increasingly environmental despoliation; some local populations are picking up on this, appropriating Western scientific and 'green' ideas to protect their interests. Squatters in Kathmandu caught up in a fierce political battle seek to promote a conservation image to counter river pollution accusations, city authorities blaming them for its sewage and garbage disposal problems. Similarly, migrant farmers in Brazilian Amazonia seek to establish themselves as environmental stewards and not the villains of forest destruction. Elsewhere in Amazonia, the indigenous Piaroa are evidencing a growing interest in traditional environmental knowledge, as Serena Heckler shows, in part to assert their identity and claim territory by

showing that they are its guardians. Furthermore, although aggressive proselyti-
sation has undermined their shamanistic tradition, they are increasingly identifying with it too as a political strategy.

The political dimension is evident in the Labrador case too, as Trudy Sable
and colleagues make clear, describing how Innu collaborators help to identify
and define project research issues in line with demands that they demonstrate to
the government their credentials as environmental guardians as a prelude to
achieving self-determination. In order to attain self-government they have to be
able to argue in, and show some understanding of, the federal government's scientific language. The reaffirmation of the Innu's tacit and practical knowledge
from a scientific perspective gives it political credence and significance, ensuring
that it features in reports passed on to politicians and policy makers, which is
central to their achieving a degree of autonomy in their own country and control over their own destinies, so long denied under colonial rule. The political
environment can seriously interfere with the dissemination of scientific research
findings via extension services, as Mariella Marzano recounts, where in Sri Lanka
complex interactions between different levels of the state bureaucracy lead to
confusion, and local people adept at manipulating their negotiations with outsiders may frustrate the efforts of extension workers as they seek to achieve other
goals. A history of partisanship and corruption has also undermined the trust
and co-operation necessary to effective extension work, and local communities
used to a paternalistic regime that issues directives are not well placed to respond
to development's current encouragement of participation. Furthermore – and
paradoxically, given current participatory rhetoric – extension arrangements
often inhibit farmer experimentation and innovation by the conditions they
impose on those who adopt their recommendations. The short-term fashion
trends that characterise development result in farmers receiving confused advice
too, promoting one thing one year and something different the next. Violent
fluctuations in market prices exacerbate this trend as farmers switch from one
crop to another in large numbers following extension advice. This brings us to
the commercial focus of the capitalist political orders that dominate and manipulate global science.

Science for Sale

The market infiltration of science has become increasingly obvious in the last
two decades with growing political pressure to fund research via partnerships
with commercial companies. The argument is that forcing closer links with commercial interests will ensure scientific research is more relevant to the demands
of the economy, and driven less by pure intellectual curiosity, which in the view
of many politicians results in work of dubious worth. What they seem to overlook is that curtailing 'blue skies' research will inhibit future breakthroughs.
History suggests that we have to allow for many blind alley projects for every

new highway discovery. We have only to consider what cranks Victorian society thought persons such as Faraday for researching electromagnetic forces, unable to see the contemporary electronic age. Furthermore, as some political commentators point out, while the profit motive that fuels capitalism might make a few wealthy, it reduces many to poverty, which is a particular concern currently in development contexts with the emphasis on poverty reduction. 'Fee-fi-fo-fum, I smell money, everyone': we have the greedy exploitation of technological advances to enrich a few, not community-focused progress. We have something to learn from political and economic arrangements elsewhere, if we believe in fairness. David Cleveland and Daniela Soleri illustrate some of these issues in their discussion of the 'economically rational farmer'. This is not the place to rehearse arguments for and against increasing links between scientific research and the market but we should note the implications for relations between local and global science. These again are largely political issues, focusing on disquiet about capitalist imperialism. They are manifest particularly in concerns currently to protect people's intellectual property rights (IPRs), to prevent unfair exploitation of their knowledge for commercial gain.

Intellectual property has become a burning issue recently, and Charles Clift takes up the argument that it is necessary to define rights legally to protect local knowledge holders. He points to some intriguing parallels that the impact of IPRs have on global and local science, arguing that pushing IPRs in both these domains could have similar deleterious consequences: for instance, undermining the collegiality central to university life as scientists become less willing to share knowledge, thus upsetting the fine balance between competition and cooperation. Patenting up-stream processes could be particularly damaging, as these often feature in future research. Financial gain has hitherto not motivated scientists so much as peer recognition, intellectual curiosity and advancing the frontiers of knowledge, which is deeply satisfying, like writing good poetry or doing beautiful painting. Also, the supposed financial benefits in applying IPRs to the university research market are largely lost in administrative costs and litigation. Clift argues that IPRs could likewise disrupt the balance of local communities, if some claim to own knowledge above others. Indeed they strike at the foundations of local knowledge, much of which is communal not private. The imposition of IPRs on local knowledge, which has flourished for millennia without such arrangements, is an aspect of the intrusion of the capitalist order into these communities (Cleveland and Murray 1997). They may disrupt established processes for generating ·new knowledge locally, as in scientific research, by upsetting its vigour, in seeking to protect and freeze it in some timeless traditional past. The idea that IPRs offer some kind of general protection to local science by assigning rights and excluding others, should not be misconstrued as protecting knowledge from extinction under the relentless onslaught of the economic and social forces of capitalism. Regarding IPRs as a solution to the protection of local knowledge is thus to misdiagnose the problem. It is the larger issues related to the penetration of the market, the loss of land and other rights that are the real problem. IPRs are not an effective way of tackling poverty and global inequalities, rather the opposite. They may thwart scientific access to knowledge for the benefit of humankind (e.g. in researching new drugs), by prompting communities to exclude researchers, or prompting them to overvalue their knowledge. The patent system is intended to provide an incentive for invention and innovation – and space for these to occur – which is quite at odds with ensuring equity in the use of local knowledge or its protection.

The political position varies from one region of the world to another, making generalisation about IPRs difficult, as some other contributors show. Gerard Bodeker argues that medical researchers must accord equal status to traditional healers when working on herbal medicines, and not allow science to dominate, as such power differentials can fatally undermine research, as illustrated with *changshan*, the Chinese anti-malarial. This demands respect, he points out, for the intellectual property rights of healers. In their contribution, Alberto Arce and Eleanor Fisher describe the legal battle over attempts to patent genetic material from quinoa, an Andean crop, by U.S. university researchers. It shows how negotiations at the knowledge interface between local and global science is complex, sometimes involving many parties. Here NGOs (non-governmental organisations) and people's organisations seek to defend farmers' interests, revealing a conflict between local and scientific goals, with serious political implications – biotechnology and genomics threatening local farmer autonomy by controlling seed supply. Elsewhere in Latin America, IPRs are of concern to many local populations, particularly in Amazonia, which is seen as biodiverse and likely to yield important new natural compounds for use in drugs, cosmetics and food. People are anxious to protect their environmental knowledge from outsiders for fear of biopiracy, as Serena Heckler relates, having learnt that their forest may contain commercially valuable products. This illustrates Charles Clift's point about IPRs possibly prompting communities to inhibit scientific research that could benefit humankind or overestimate the commercial value of their resources.

Dangerous Science?

The idea is not that the small local knowledge stone should knock Goliath science over, an improbable, even ridiculous prospect. It is that we should create space for others' ideas. This is necessary not only because it should continue to add to global science's awesome fund of knowledge, but also because it might help us to manage this knowledge more effectively for the planet and humankind. It is becoming increasingly evident that our astounding scientifically informed technological advances are coming at considerable and possibly unsustainable costs. There is growing discontent with science in many sections of our society – from light-green conservationists to deep-green eco-warriors, purple establishment Church of England to rainbow millenarian cults, and pink neo-left politicians to dark-red socialists – which chime in with calls for us to listen to local voices (Milton 1993). This is not to suggest that other traditions are wiser than science. In anthropology we have learnt to tread the middle road between thinking that all traditional tribal ways are inherently sustainable and environmentally sound (Ellen 1986), and the reverse that all peasants are ignorant and demand development, or worse civilisation. We are all fallible humans. But we think, of course, that we have much to learn from all people; this is a *sine qua non* of the discipline.

Many within development circles, with their short historical perspective and rapid idea fashion cycles, date concerns with sustainability to the so-called 'Brundtland report' (World Commission on Environment and Development 1987), whereas demographers routinely trace them back to Malthus (1798) and the early nineteenth century, although there is evidence of population concerns much earlier, for instance, 1600 BCE Babylonian tablets refer to problems of overpopulation (Cohen 1995). There are concerns about human population levels and global carrying capacity, which come and go in prominence. There are worries about damage to the environment with ozone holes and global warming, climatic changes and pollution. How irresponsible can we be before a catastrophe occurs? Take the generation of nuclear power, which creates in plutonium a deadly radioactive waste product with a half-life of 10,000 years, that we are dumping in subterranean silos in northern England where 10,000 years ago there was an ice age and glaciers carved valleys through the rock. No fear (just yet) of global warming of glaciers pushing canisters of plutonium into the Home Counties but what a liability we are passing onto future generations. We are increasingly asking ourselves if science is intervening in nature, and indirectly social arrangements, in ways that we lack the wisdom to manage, meddling with life itself in test-tube reproduction, genetic modification of crops, and so on, not to mention weapons of unimaginable destructive capacity. There is anxiety about the social implications of technologically driven change, with evidence of breakdown and alienation, isolation and hostility. We are entering dangerous new territory with the erosion of the family, which for millennia has been the cornerstone of social order for cultures around the world. We have good reason to be expressing concerns about sustainability, and it is not only non-scientists who are doing so. Scientists are too, as evidenced by the 'sustainable science' theme of the British Association Festival in 2003 at which this book came into being.

These worries suggest that scientific knowledge is dangerously partial, that there are some fundamental gaps in its worldview and that its dominance is unhealthy. What is problematic about the scientific approach? There have been several critiques, as the essays in this volume indicate. The contribution of David Cleveland and Daniela Soleri tackles the issue of sustainability head on. They open by arguing that we can only define sustainability in agricultural contexts subjectively, based on the values that we bring to the definition. But once we have agreed a definition we can objectively assess indicators of sustainability. The authors point out that farmers' knowledge has increasingly featured in debates about sustainable agriculture, but that it tends to be defined deductively according to the assumptions that inform the definition of sustainability, whereas they argue that we should empirically investigate farmers' knowledge to see how it complies with sustainable goals. Furthermore, they argue that definitions of farmer knowledge and scientific knowledge often determine the roles of farmers and scientists in sustainable agriculture development. They elaborate on this thesis by proposing four approaches to local farmer knowledge, which they call the 'economically rational farmer', the 'socially rational farmer', the 'ecologically rational farmer' and the 'complex farmer' approaches. They take each in turn and give the definition of sustainable agriculture they entail, followed by definitions of local farmer knowledge and global scientific knowledge to which they lead, before describing the different roles for farmers, and the natural and social scientists in agricultural development that these approaches stipulate.

Bob Rhoades and Virginia Nazarea further address the issue of sustainability. They use modified folktales and panoramic photographs in methodologically innovative participatory contexts to probe people's views of landscape changes, to compare with the work of scientists researching models of conservation. They reveal an emphasis locally on community and livelihood concerns, notably the need for economic development to improve the standard of living for people who in the past have had a raw deal. Deforestation is not a prominent concern as it is with the scientists whose values and assumptions about conservation inform their modelling predictions (which depend on the parameters they feed in, such as monitoring forest loss, land degradation and so on). It is not that the scientists are wrong to focus on conservation but that any attempts to persuade local people to buy into the need to protect the ecologically important zones found in their region are unlikely to succeed if they do not also address the issues those local people think important. It is not the familiar scenario of sustainable local practices versus unsustainable science-technology but of formulating biodiversity conservation policies in ways that relate to local concerns.

A belief in sustainability draws on deeply held Aboriginal values, as Ben Smith makes clear in his contribution, relating how people blocked plans to log their forest. Heirs to one of the most conservative cultures known – according to the archaeological record Aboriginal lifeways remained largely unchanged for generations – where existence evidences a circularity and demands repeated contemporary validation of creation-time events, sustainability is beyond question, being central to life as they know it. In my own contribution I contrast circular ideas of being, which reflect more sustainable worldviews, with the linear perspective of science. The way in which we count informs these views. The limitless counting scheme that characterises the mathematics of science, while it underpins many startling discoveries and innovations, intimates an unsustainable perspective, as graphically illustrated by the concepts of infinity and zero. This contrasts with the finite perspective of counting systems found elsewhere, such as in New Guinea. Focusing on the computational limitations of such schemes prompts us to write them off, which is unfortunate. The finite standpoint informing them imbues sustainable views, in contrast to the Euro-Asian scheme as increasingly evident with the current harnessing of science to the market economy with its nostrums for endless growth while we inhabit a finite planet. Those who count less may have something to teach us about sustainability.

It is not our intention in local knowledge work to enrage, after Jack of the Beanstalk fame, the science giant but to work with it to promote sustainable and appropriate interventions. We argue that we need to draw on other cultural perceptions to achieve balance. We have to acknowledge the social positioning of knowledge – that understanding is culturally embedded, and not value free (including that of science). We find ourselves on the horns of a dilemma (Dunbar 1995; Peat 2002). Some argue that it is pointless to talk about controlling scientific research because inquisitive humans will always engage in it because nature exists out there and waits for us to unravel her codes. The genie is already out of the bottle. But are we ready to play God with the knowledge? All the evidence suggests not. The worry is that we do not have the wisdom to handle some of the knowledge, such as designer genetics for offspring, horrendous weapons of mass destruction in the hands of fallible humans, whether politicians or terrorists. Perhaps others can help us to find a morality that can assist us to ensure that we manage this knowledge better than market capitalistic democracies appear able to do. But there are no straightforward answers. Some other codes of morality would seek to control scientific enquiry, keep us living in virtual medieval ignorance, implying feudal political control. We are arguing that we need to draw on the full range of the human heritage as we seek ways forwards in the future that might benefit all humankind and ensure the continued well-being of the planet we inhabit.

Notes

1.   The implication of these examples is not that local people always get it right. There are many instances where people make mistakes; for example, refugees often find that their knowledge ill equips them to manage in the regions to which they are displaced. See Dunbar (1995: 47–57) for some further ethnographic examples.

2.   Part of the U.S. Government's Sustainable Agriculture and Natural Resource Management Project (SANREM) response to Agenda 21 of the Rio Earth Summit.

3.   The project involves the Innu Nation of Labrador, Environment Canada (a federal government department) and the Gorsebrook Research Institute (St Mary's University, Nova Scotia).

References

Antweiler, C. 1998. 'Local Knowledge and Local Knowing: an Anthropological Analysis of Contested "Cultural Products" in the Context of Development', *Anthropos* 93: 469–94.

Atran, S. 1990. *Cognitive Foundations of Natural History: Toward an Anthropology of Science.* Cambridge: Cambridge University Press.

Chalmers, A.F. 1999. *What Is This Thing Called Science?* St Lucia, Brisbane: University of Queensland Press.

Cleveland, D.A. 2001. 'Is Plant Breeding Science Objective Truth or Social Construction? The Case of Yield Stability', *Agriculture and Human Values* 18(3): 251–70.

Cleveland, D.A. and S.C. Murray 1997. 'The World's Crop Genetic Resources and the Rights of Indigenous Farmers', *Current Anthropology* 38: 477–515.

Cleveland, D.A. and D. Soleri (eds) 2002. *Farmers, Scientists and Plant Breeding: Integrating Knowledge and Practice.* Wallingford: CABI Publishing.

Cohen, J.E. 1995. *How Many People Can the Earth Support?* New York: W.W. Norton & Company.

Conklin, H.C. 1957. 'Hanunoo Agriculture: a Report on an Integral System of Shifting Cultivation in the Philippines, *F.A.O. Forestry Development Paper* No. 12. Rome: Food and Agriculture Organization.

Darwin, C. 1845 *Journal of Researches into the Natural History and Geology of the Countries Visited during the Voyage f HMS Beagle around the World: under the Command of Capt Fitzroy.* London: John Murray.

Deveraux, A.P. and E.M. Evans 2004. 'Religious Belief, Scientific Expertise, and Folk Ecology', *Journal of Cognition and Culture* 4(3 & 4): 485–524.

DeWalt, B. 1994. 'Using Indigenous Knowledge to Improve Agriculture and Natural Resource Management', *Human Organisation,* 53(2): 123–31.

Dickson, D. 1999. 'ICSU Seeks to Classify "Traditional Knowledge"', *Nature* 401: 631.

Dunbar, R.I.M. 1995. *The Trouble with Science.* London: Faber and Faber.

Ellen, R.F. 1986. 'What Black Elk Left Unsaid: on the Illusory Images of Green Primitivism', *Anthropology Today* 2: 8–12.

———— 2004. 'From Ethno-Science to Science, or "What the Indigenous Knowledge Debate Tells Us about How Scientists Define Their Project"', *Journal of Cognition and Culture* 4(3 & 4): 409–50.

Ellen, R. and H. Harris 2000. 'Introduction', in *Indigenous Environmental Knowledge and Its Transformations,* (eds), R. Ellen, P. Parkes and A. Bicker, pp. 1–33. Amsterdam: Harwood.

Fairhead, J. and M. Leach 1996. *Misreading the African Landscape: Society and Ecology in a Forest-Savanna Mosaic.* Cambridge: Cambridge University Press.

Fisher, H.J. 2001. *Faraday's Experimental Researches in Electricity: Guide to a First Reading.* Santa Fe: Green Lion Press.

Forsyth, T. 1996. 'Science, Myth and Knowledge: Testing Himalayan Environmental Degradation in Thailand', *Geoforum* 27(3): 375–93.

Frake, C.O. 1983. 'Did Literacy Cause the Great Cognitive Divide?' *American Ethnologist* 10: 368–71.

Gooding, D.C. and F.A.J.L. James (eds) 1985. *Faraday Rediscovered: Essays on the Work and Life of Michael Faraday, 1791–1867.* London: Macmillan.

Ingold, T. 2004. 'Introduction: Anthropology after Darwin', *Social Anthropology* 12(2): 177–79.

ISSJ 2002. 'Special Issue on Indigenous Knowledge', *International Social Science Journal* No. 173.

Jevons, F.B. 1900. 'Savage Science', *Proceedings of the University of Durham Philosophical Society* 1 (1896–1900): 11–25.

Knorr-Cetina. K. 1999. *Epistemic Cultures: How the Sciences Make Knowledge.* Cambridge, MA: Harvard University Press.

Lansing S.J. 1991. *Priests and Programmers: Technologies of Power in the Engineered Landscape of Bali.* Princeton, NJ: Princeton University Press.

Latour, B. 1987. *Science in Action: How to Follow Scientists and Engineers through Society.* Cambridge, MA: Harvard University Press.

———— 1999. *Pandora's Hope: Essays on the Reality of Science Studies.* Cambridge, MA: Harvard University Press.

Latour, B. and S. Woolgar 1986. *Laboratory Life: the Construction of Scientific Facts.* Princeton, NJ: Princeton University Press.

Leach, M. and R. Mearns. (eds) 1996. *The Lie of the Land: Challenging Received Wisdom on the African Environment.* London: Routledge.

Malinowski, B. 1922. *Argonauts of the Western Pacific.* London: Routledge and Kegan Paul.

Malthus, T.R. 1798. *An Essay on the Principle of Population, as It Affects the Future Improvement of Society.* London: printed for J. Johnson.

Martin, G. 2002. 'The Innovative Wisdom Process'. Unpublished notes on Global Diversity Foundation seminar series.

Milton, K. (ed) 1993. *Environmentalism: the View from Anthropology.* London: Routledge (ASA Monograph Series No. 32).

Nader, L. (ed) 1996. *Naked Science: Anthropological Enquiry into Boundaries, Power and Knowledge.* New York: Routledge.

Nasr, S.H. 1976. *Islamic Science.* Westerham: World of Islam Publishing Co.

Peat, F.D. 2002. *From Certainty to Uncertainty: the Story of Science and Ideas in the Twentieth Century.* Washington D.C.: Joseph Henry Press.

Popper, K. 1959. *The logic of Scientific Discovery.* London: Hutchinson.

Purcell, T.W. 1998. 'Indigenous Knowledge and Applied Anthropology: Questions of Definition and Direction', *Human Organization* 57(3): 258–72.

Puri, K. 2001. 'Legal Protection of Traditional Knowledge and Cultural Expressions of Indigenous Peoples'. Lecture given at 'Innovation, Creation and New Economic Forms' conference, Corpus Christi College, Cambridge 13–15 December 2001.

Richards, P. 1985. *Indigenous Agricultural Revolution: Ecology and Food-crop Farming in West Africa*. London: Hutchinson.

———— 1986 *Coping with Hunger: Hazard and Experiment in a West African Rice-farming System*. London: Allen & Unwin.

Sillitoe, P. 1996. *A Place Against Time: Land and Environment in the Papua New Guinea Highlands*. Amsterdam: Harwood Academic.

———— 1998. 'The Development of Indigenous Knowledge. A New Applied Anthropology', *Current Anthropology* 39(2): 223–52.

———— 2002. 'Globalizing Indigenous Knowledge', in *'Participating In Development': Approaches to Indigenous Knowledge*, (eds), Paul Sillitoe, Alan Bicker and Johan Pottier, pp. 108–38. London Routledge.

Sillitoe, P., A. Bicker and J. Pottier (eds) 2002. *Participating in Development': Approaches to Indigenous Knowledge*. London: Routledge (ASA Monograph Series No. 39).

Spranzi, M. 2004. 'Galileo and the Mountains of the Moon: Analogical Reasoning, Models and Metaphors in the Scientific Discovery', *Journal of Cognition and Culture* 4(3 & 4): 451–83.

Turnbull, D. 2000. *Masons, Tricksters and Cartographers: Comparative Studies in the Sociology of Scientific and Indigenous Knowledge*. Amsterdam: Harwood Academic.

Verran, H. 2002. A Postcolonial Moment in Science Studies: Alternative Firing Regimes of Environmental Scientists and Aboriginal Landowners. *Social Studies of Science* 32(5 & 6): 729–62.

Whewell, W. 1989. *Theory of Scientific Method*, (ed.) R.E. Butts. Indianapolis: Hackett Publishing Co.

World Commission on Environment and Development. 1987. *Our Common Future: Report of World Commission on Environment and Development*. Oxford: Oxford University Press.

2 Traditional Medical Knowledge and Twenty-first Century Healthcare: the Interface between Indigenous and Modern Science

Gerard Bodeker

In the assessment of the World Health Organization (WHO), the majority of the population of most developing countries regularly use and rely on traditional medicine for their everyday healthcare needs. At the same time, policy and regulation in support of this social and public health reality are still in an early stage of formation in most countries (Bodeker et al. 2005).

In response to a call from member countries to give greater emphasis to traditional medicine policy development, the World Health Organization's Traditional Medicines Strategy 2002–2005 was formed to focus on four areas identified as requiring action if the potential of traditional, complimentary and alternative medicine (TCAM) to play a role in public health is to be maximised. These are: policy; safety, efficacy and quality; access; and rational use. Underpinning the WHO move to integrate traditional medicine into national healthcare in its member states is a call for an evidence-based approach to traditional and complementary medicine. This, in its strictest sense, means the application of randomised controlled clinical trial methodology to the evaluation of herbal and other therapeutic modalities used in these systems. In a more general sense it means the development of a body of knowledge based on Western methods of empirical investigation, including laboratory and animal studies on the safety and efficacy of traditional herbal and other therapeutic modalities. Indigenous groups and traditional medicine spokespeople have pointed out that the requirement for evidence to be gathered according to Western understandings of traditional knowledge (TK) can have the effect of reducing long-held theoretical constructs of the body, disease pathogenesis and therapeutic modalities to testable elements that bear little or no relation to the original constructs on which theory and practice have been based throughout their long histories. This debate is one that is central to the interface between Western biomedicine and traditional medical systems and this chapter will attempt to address some of the central issues in terms of both the preservation of TK standards and the development of tradition-based healthcare in a contemporary setting.

Traditional Medicine – WHO Definitions

Traditional, i.e. indigenous, medicine has received prominent attention in the past decade as pharmaceutical companies have come to view traditional medicines as possible sources of potent molecules that may be replicated synthetically and patented as new and profitable pharmaceutical products. It has also been the subject of new interest by health policy makers as a field to be developed in the context of providing equitable healthcare coverage in developing countries (Bodeker 2001).

The World Health Organization (WHO) has offered the following conceptualisation of traditional medicine:

> On the basis of a community's or a country's culture, history and beliefs, traditional medicine came into being long before the development and spread of western medicine that originated in Europe after the development of modern science and technology. The knowledge of traditional medicine is often passed on verbally from generation to generation. Nevertheless, in some cases a sophisticated theory and system is involved. (World Health Organization 1978)

Intrinsic to this definition is the notion of community or cultural ownership of their traditional medical heritage – a perspective with bearing on the now-contentious area of intellectual property rights (IPR) and traditional knowledge (TK), which will be addressed later in this chapter.

Cosmology

An essential feature of traditional health systems is that they are based in cosmologies that take into account mental, social, spiritual, physical and ecological dimensions of health and well-being. A fundamental concept found in many systems is that of balance – the balance between mind and body, between different dimensions of individual bodily functioning and need, between individual and community, individual/community and environment, and individual and the universe. The breaking of this interconnectedness of life is a fundamental source of dis-ease, which can progress to stages of illness and epidemic. Treatments, therefore, are designed not only to address the locus of the disease but also to restore a state of systemic balance to the individual and his or her inner and outer environment (Bodeker 1996).

In earlier policy statements, the World Health Organization has referred to the world's traditional health systems as 'holistic' – 'i.e. that of viewing man in his totality within a wide ecological spectrum, and of emphasizing the view that ill health or disease is brought about by an imbalance, or disequilibrium, of man in his total ecological system and not only by the causative agent and pathogenic evolution (WHO 1978).

The Asian medical traditions, particularly the systems of India and China which have been the source of many of the medical traditions of other countries in the region, are based in theoretical frameworks and methods of discovery that are considered by their proponents to be truly scientific approaches. Underlying the approach is the view that the observer is of central importance in the process of investigation and that subjective modes of discovery, such as insight and intuition, are primary tools for advancement of knowledge. This view reflects a phenomenological position akin to both that of William James and his colleagues in the late nineteenth century and a more fundamental position that the universe is not inert but rather is imbued with – and is the source of – consciousness.

In the Vedic tradition of India, the source from which the Ayurvedic medical system derives (*Ayus* = life, *Veda* = knowledge), consciousness is considered the basis of all material existence:

The infinite consciousness alone is the reality, ever awake and enlightened.

… wind comes into being, though that wind is nothing but pure consciousness.

Within the atomic space of consciousness there exist all the experiences, even as within a drop of honey there are the subtle essences of flowers, leaves and fruits.

… even what is inert is pure consciousness.

It is pure consciousness alone that appears as this earth. (Venkatesananda 1984)

Within this framework, consciousness is of primary significance and matter is deemed secondary. Accordingly, Ayurvedic medical treatment, when practised according to the high traditions of Ayurveda, will first address the spiritual and mental state of the individual – through meditation, intellectual understanding of the problem, behavioural and lifestyle advice, etc. – and then address the physical problem by means of diet, medicine and other therapeutic modalities (Sharma and Clark 1998).

The seminal text of traditional Chinese medicine, the *Shen Nong Ben Cao*, or Yellow Emperor's Classic, dated around AD 200 (Unschuld 1992), also details a cosmology in which matter is secondary to ethereal dimensions of existence. The cosmos is described as composed of ethers of heaven and earth, which are yang – with the attributes of bright, light and male – and *yin*, with the attributes of dark, heavy and female. The universe contains phenomena created by the dynamic action upon yin and yang of the Five Agents (*wu-hsing*), the elements water, fire, metal, earth and wood, which mutually create and destroy each other. The concept of *qi* refers to subtle energy or life force and is further described in categories which govern nourishment, defence systems, flow of energy, physiological functioning of organs, respiration and circulation. The relationship among these phenomena in their natural state is one of balance and harmony.

Such a view reflects the existence of a coherent theoretical framework underlying many traditional health systems. In addition to theory, these systems use taxonomic frameworks for classifying diseases and the medicinal plants used to treat them. Indeed, traditional health systems have organised frameworks for classifying plants, animals, landscapes and climatic conditions in relation to their effects on health and disease. These taxonomies have much in common with one another and represent a culturally relevant empirical framework for assessing medicinal plant biodiversity.

Nature is also valued as having a vital reality beyond the observable. Harnessing this reality is the essence of restoring balance – ease from dis-ease – be it understood in terms of the energy described as *qi* by traditional Chinese medical practitioners, or as *prana* or consciousness by the Ayurvedic exponents, or the various characterisations of the spiritual realm found in African, Pacific and South American health traditions.

Observations expressed in spiritual terms may have strong empirical validity, as can be seen from this perspective in the ancient Vedic text, the *Shrimad Bhagavatam*: 'The sun is the soul of the deities, men, beasts, reptiles, creepers, and seeds' (Rishi Sukadeva, XXII, pp. 508–9, *Shrimad Bhagavatam*). Honouring and using, revering and understanding, harvesting and conserving – these may not represent the polarities that, from a secular vantage point, they may appear to be. Their coexistence may be more than the tolerance of inconsistency or the capacity to live with paradox. It is, conceivably, the expression of an appreciation of a deeper level of unification between the different dimensions of life, where paradox is seen more as the product of superficial or fragmented perception than as an accurate rendition of a meaningless universe.

The cosmological dimension of traditional knowledge systems, specifically as they pertain to health in this context, is both of core significance in understanding their therapeutic modalities and outcomes and yet at the same time poses a challenge to biomedical research to capture and address both the outcomes and the theoretical assumptions on the mechanisms by which therapies have their effect.

Recent work to bridge the seeming divide between non-Western theoretical models of physiology and treatment has offered explanations from theoretical physics to explain the Ayurvedic tri-*dosha* theory. Hankey (2001) has argued that the three *doshas* of Ayurveda and their five respective sub-*doshas* are related to the modern scientific framework of systems theory, phase transitions, and irreversible thermodynamics. Hankey has argued that the three-*dosha* theory represents far more general biological concepts than the neuroendocrinology of their functioning might imply. It is proposed that they express universal concepts applicable across living organisms – control structures governing living systems. The description of varying states of health and disease given in Ayurvedic aetiology is related to the format of phase transitions in irreversible thermodynamics.

Linking advances in the physical theory of critical phenomena, Hankey (2004) has proposed that physical instabilities result in fluctuations and that the quantum properties of these fluctuations can be applied to regulatory control mechanisms in living organisms with promising results. From this perspective, many aspects of the energy theories (qi, prana) of traditional medicine can be scientifically modelled, in agreement with existing theoretical concepts, such as the existence of macroscopic quantum coherence in living systems.

Taxonomy

The Ayurvedic system of India identifies organising principles (*doshas*) which underlie physiological make-up, disease and biodiversity and uses six tastes to determine the doshic properties of plants. The Maya of Mexico have a framework which uses taste, or an organoleptic approach, to assess the suitability of plants for specific categories of conditions. The Barasana people of northwest Amazonia and the Aka pygmies of central Africa classify food and medicinal plants according to their hot or cold properties, as is also the case in traditional Chinese medicine, and also use taste categories bearing similarities to those in other traditions. Brent and Eloise Anne Berlin (1996), distinguished medical anthropologists, have noted that the taxonomies of traditional medical systems as diverse as Mexican and Chinese have more in common with one another than any of them have with the Western medical framework. Such commonality of culturally unconnected theories and systems raises the question of whether traditional health systems, through an empirical and scientific approach to disease and *materia medica*, may be tapping into a common set of natural principles or laws as yet to be determined by Western scientific investigations.

Traditional health knowledge extends to an appreciation of both the material and non-material properties of plants, animals and minerals. Such knowledge systems include concepts of both the sacred and the empirical, frameworks for understanding health and healing, assumptions of cosmos and causality and taxonomies which address a perceived order in Nature. Traditional classificatory systems range in their scope from the cosmological to the particular in addressing the physiological make-up of individuals and the specific categories of *materia medica* – the materials used for therapeutic purposes in traditional health systems – needed to enhance health and well-being.

While having a theoretical foundation in the non-material realms of existence, indigenous medical traditions may draw on plant and animal taxonomies which have much in common with one another – and which diverge from Western classificatory frameworks and assumptions.

The proposition by Brent Berlin in the early 1970s that consistent categories exist within indigenous systems of classification – ethnotaxonomies – has served as a catalyst to opening up the study of traditional naming systems (Berlin 1973). The result has been new insights into the effects and adaptiveness of indigenous applications of plants.

Andean shamans of the Sibundoy Valley, characterising the plants in their own gardens, exhibited a greater degree of discrimination and a more diverse range of information in the ethonobotanical categories that they employed than the categories used by Western botanists. The gardens of the shamans reflected their knowledge of ecosystem management as well as serving as symbolic reconstructions of the web of relationships between plants, humanity and the cosmos (Pinzon and Garay 1990).

In northern Brazil, Yanomami Indian herbalists, faced with a severe epidemic of malaria resulting from the gold rush, adopted, from necessity, an empirical approach to identifying those Amazonian plants which might have an anti-malarial effect. In examining the anti-malarial effects of plants conventionally used for controlling other forms of fever, one major criterion in the identification of plants that had an anti-malarial effect was the characteristic of bitterness (Milliken 1997).

Bitterness is also the taxonomic category used in Ayurveda and traditional Chinese medicine for selecting plants which have anti-pyretic effects. A bitter taste serves as the basis for selecting the plants used in many Indian and Chinese traditional anti-malarials.

Indeed, Ayurveda uses six tastes as the basis for classifying the principles of food and medicines. They are: sweet, sour, salty, bitter, pungent and astringent. A basic principle of the Ayurvedic approach to nutrition is that through the sense of taste the human physiology has a latent capability to spontaneously determine the dietary and nutritional needs appropriate to individual need, if the physiological functioning of the individual is balanced. If it is not, the physician determines the nature of the imbalance through diagnostic means, including pulse diagnosis, and recommends the use of those plants (foods and medicine) which will provide balance.

In Ayurveda, individual constitutions are classified according to these three basic organising principles or *doshas*. In diagnosing a disease, an Ayurvedic physician will first determine the *doshas* which constitute the psychophysiological type of the patient. The Ayurvedic physician asks not, '*What disease* is the patient presenting with?', but '*What type of person* is presenting with the disease?' According to the assessment of the individual and his/her state of imbalance, the physician will select treatments designed to restore balance to the patient. The choice of plants and other medicinal materials is based on their capacity to influence the *doshas*.

The classic Ayurvedic text, *Caraka Samhita*, which is between 1,500 and 2,000 years old, notes that while folk knowledge exists regarding the form and effects of medicinal plants, a higher form of knowledge is that pertaining to the principles governing correct application of the plants to human health. 'One who knows the principles governing their correct application in consonance with the place, time and individual variation, should be regarded as the best physician' (*Caraka Samhita*, Vol.1, p. 59).

In the preparation of plants, traditional Asian pharmacologies emphasise a principle of synergistic activity among the components of plant ingredients of herbal mixtures. This assumes that, just as the body is designed to extract multiple components from food, it is also designed to do the same from medicinal plant materials. Traditional medicines typically use more than a single plant. Complex mixtures of plants form the basis of prescriptions and these are frequently prepared through a process which may include drying, crushing, heating, boiling, even reducing to a form of ash. Consequently, a chemical process is involved which transforms the chemical structure of the plant materials and produces a set of compounds which may be different from those contained in each of the individual plants in the prescription.

Ecosystems interact with the human physiology and are associated with different health outcomes. Within the Vedic classificatory system, ecosystems are also classified into the three main *doshic* categories. Soil types are classified into five additional categories: black, red, white, yellow and blue, showing yet again the degree of differentiation of the Vedic classificatory system.

While starting from a cosmology which elevates the non-material to a level of superiority over, or precedence to, the material, many traditional health systems have a pragmatic, empirical and analytic basis to them. Diseases and treatments are classified based on the specific properties of the plant or animal products which they contain. Individual psychophysiological characterisitics can be categorised to a fine degree, allowing for individualised treatments. This shares much with the contemporary biomedical orientation to the understanding of symptoms, aetiology, disease types and categories of individual differences in disease. The non-Western view of nature and its role in healing is also characterised by this analytic approach. Berlin and Berlin (1996) have noted that anthropology, in focusing on healing rituals and beliefs, has made the mistake of overlooking the genuinely empirical, pragmatic and naturalistic dimensions of traditional health systems in favour of phenomena which inspire wonder in the observer.

Interface of Western and Traditional Science: Pharmacological and Clinical Research

It is important that clinical evaluation of traditional medicines be conducted with consideration of the concepts and practices of traditional medical systems. It has been noted (Chaudhury 2002) that the traditional medical diagnostic framework highlights differences in response pattern of different constitutional types of patients to a common drug. Ayurvedic medicine, for example, delineates three major constitutional types, known as *doshas* – *vata*, *pitta* and *kapha* (Sharma and Clark 1998). The Ayurvedic medicine, *kanaka*, is considered effective in patients with the Ayurvedic body type of *kapha* who have bronchial asthma, but for patients of the *pitta* type, it is predicted to cause congestion. Such considerations should be taken into account in patient assignment and randomisation can be done, for example, taking into account traditional diagnostic categories.

The Indian Council of Medical Research has developed two herbal drugs – a contraceptive and a treatment for diabetes mellitus –using randomised controlled clinical trials (RCTs) guided by the clinical assumptions and diagnostic categories of India's Ayurvedic medical system. This requires that traditional medical practitioners be part of clinical research teams and that their perspectives be given serious and respectful consideration in designing appropriate RCTs of their medicines.

Traditional medicines, based on the use of whole plants with multiple ingredients or of complex mixtures of plant materials, could be argued to constitute combination therapies that may combat the development of resistance to antimalarial therapy. David Warrell (1997), Nuffield Professor of Medicine at Oxford University, has noted that: 'Whereas testing of individual compounds may lead to identification of the sole or major active component, possible synergism among the different ingredients or the special effects of the mode of preparation may be lost or obscured.' Williamson (2001) has reviewed the evidence for the occurrence of synergy in plant-based medicines. She notes that: 'Synergistic interactions are of vital importance in phytomedicines, to explain difficulties in always isolating a single active ingredient, and explain the efficacy of apparently low doses of active constituents in a herbal product. This concept, that a whole or partially purified extract of a plant offers advantages over a single isolated ingredient, also underpins the philosophy of herbal medicine. Evidence to support the occurrence of synergy within phytomedicines is now accumulating.'

Changshan – a Cautionary Tale for Reductionism

Changshan, a 5,000-year-old traditional Chinese febrifuge, was developed by the Chinese government early in the twentieth century as a national anti-malarial drug. Despite its efficacy as an anti-malarial and millennia of use as an effective febrifuge for hot-cold fevers, *changshan* was abandoned as an anti-malarial drug less than three decades after the modern R&D (research and development) process began. The process by which *Changshan* was 'modernised', the associated problems with compliance resulting from side effects not present in the traditional formulation, and the subsequent abandonment of a viable anti-malarial, constitute an important historical case study in the development of anti-malarial drugs from traditional medicine.

The earliest record of *changshan* comes from the *Shen Nong Ben Cao*, dated around AD 200 (Unschuld 1992). The *Shen Nong Ben Cao* lists the known pharmacopoeia of the day and reports *changshan* to be helpful in treating fevers. Towards the end of the sixth century AD, one of the more than one hundred prescriptions carved into the walls of the Buddhist caves of Longmen (also known as Dragon's Gate), south of Luoyang in Henan province, contains Ge Hong's use of *Dichroa febrifuga*.

There are several versions of *changshan* recorded in classical Chinese texts – in each case it is a complex mixture containing as a central ingredient the plant now known to be *Dichroa febrifuga Lour*. While there have indeed existed several versions of *changshan*, as a rule this name refers to the root of the plant *D. febrifuga*, rather than to a complex mixture.

D. febrifuga is a small genus of twelve species of evergreen shrubs in the saxifrage family, Saxifragaceae. *D. febrifuga* ranges from India and Nepal eastwards to southern China and into southeast Asia, growing at forest edges at altitudes of between 900 m to 2,400 m. The plant resembles a small-leaved Hydrangea and its flowers, blue in colour, are formed at the tips of the shoots from summer onwards

The formula that was the basis for research and development in early twentieth-century China came from a prescription circulated in a newspaper, reportedly from classical sources (Guoyao Yanjiushi 1944). This formulation is a complex mixture of different species consisting of seven ingredients: *Dichroa febrifuga* Lour., areca nut (*Areca catechu* L.), tortoise shell (*Carapax Amydae Sinensis*), liquorice (*Glycyrrhiza glabra* L.), black plum (*Prunus mume* (Sieb.) Sieb. et Zucc.), red jujube (*Ziziphus jujuba* Mill.) and raw ginger (*Zingiber officinale* Rosc.).

While *Dichroa febrifuga* was used in combination with several other plants in the classical Chinese formulations, in the Himalayas a decoction of the leaves, bark and roots of *Dichroa febrifuga* has been taken traditionally for fever. Although the plant does not appear to have a Sanskrit name and is not described in classical Ayurvedic texts, it is known in folk medical traditions by the Hindi name '*basak*' (Foundation for the Revitalisation of Local Health Traditions, India, personal communication, 2002). In the preparation of the medicine, caution is exercised, as the fresh juices are known to be emetic. These effects are reported elsewhere, where *Dichroa febrifuga* is listed as a known emetic (http://www.floridaplants.com/Med/emetic.htm).

The choice of plant ingredients in the formula including *changshan* was guided by classical Chinese medical theory (see the Yellow Emperor's Classic, c. 300 BCE trans. Ni 1995), which would favour the selection of multiple – rather than single – ingredients that were known to be suited to the type of fever and the type of individual to be treated.

By contrast, the Western medical search for new antimalarials from plants has followed a different track. Typically, the search has begun with a clue from the indigenous use of plants in the management of fevers, such as, for example, the use of *changshan* found in or reported from ancient records. Often, no further action is taken and the traditional medicine is simply not developed or applied within a modern medical context. However, when the traditional medicine has been investigated, a subsequent step in development is for reports to be gathered from patients about fevers being relieved by the medicine. Early non-indigenous accounts of use can be a source of hypothesis generation as well, such as travellers' tales of miraculous cures. If data from these sources indicates that a plant or plant mixture seems to be producing results that may show potential for malaria, a path of scientific investigation begins. Botanists work to identify the correct species that are described in classical literature or in folk oral traditions, as some plants may be used interchangeably due to regional variations in plant availability and in the use of a single common name for different species. Experimental studies are conducted to screen plant extracts for anti-plasmodial activity, and then to isolate possible 'active ingredients.'

Clearly, the two paradigms are so different from one another that it is easy to understand why each might seem incompatible with the other.

In the case of *changshan*, patient reports on self-medication led to clinical investigations which showed *Dichroa febrifuga* to be active (Chen Fangshi 1944). After this had been established, the distinguished Cambridge scholar of Chinese medicine, Joseph Needham, sent the original complex traditional herbal mixture to K.K. Chen's research group in Eli Lilly & Company (Henderson et al. 1948), and to another American group for animal experiments (Tonkin and Work 1945). When these U.S. animal experiments confirmed the anti-malarial efficacy of *Dichroa febrifuga*, some samples of the same herbal drug were sent to the University of London for pharmacognostic research (Fairbairn and Lou 1950). The other Chinese governmental laboratory that pioneered anti-malarial drugs also requested an aqueous extract of *changshan* from the Research Laboratory of National Drugs.

Isolated Active Ingredient or Synergistic Activity within a Complex Mixture?

Febrifugines, alkaloids derived from *Dichroa febrifuga*, were analysed and synthesised during the Second World War in a programme to protect American forces from malaria in the Pacific and other tropical campaigns (Koepfli 1947). Despite advances in understanding the chemistry of *changshan*, the chemists engaged in the research were unable to separate out the nausea that the drugs produced, and the result was that the *changshan*-derived anti-malarials did not prove viable (http://www.chamisamesa.net/drughist.html).

Despite the promising experimental work with *D. febrifuga* and its derivatives and the promising clinical findings from the exploratory research within China, the well-known emetic effect of *D. febrifuga*, in the absence of the accompanying plants used classically to offset nausea, was to prove a central factor in *changshan* not being adopted for widespread contemporary use as an antimalarial. Patients simply would not take it.

Evidently, the vomiting and nausea produced by D. febrifuga was intensified when the 'active' alkaloids were used in isolation from the other chemical constituents of the plant. These effects were further compounded by the isolation of *D. febrifuga* itself from the companion ingredients used in the classical formulation of *changshan*.

Polypharmacy as a Means to Offset Nausea Induced by D. febrifuga

In the case of *changshan*, it is not resistance that has been the problem leading to disuse of the drug, as we have seen, but nausea and vomiting associated with the known emetic effects of febrifugine. Not surprisingly, however, at least three of the other ingredients in the classical formulation studied in China in the mid-twentieth century would seem to be candidates for offsetting the emetic properties of *changshan*. They are *Zingiber officinale* (ginger), *Glycyrrhiza glabra* (liquorice) and *Areca catechu* (betel nut).

Zingiber officinale – ginger – is an ingredient in the classical formula, and has been found through a large body of experimental and clinical research to be effective against nausea.

Glycyrrhiza glabra (Leguminosae) – liquorice – has been the basis for an aqueous solution in which *Dichroa febrifuga* has been soaked, in a classical Chinese method, to remove the nausea-inducing effects. It has been found to reduce the side-effects, including nausea, resulting from chemotherapy in postoperative breast cancer patients (Akimoto et al. 1986). Liquorice is also known to have synergistic effects with other ingredients of plant-based medicines (Williamson 2001) and may serve such a role in facilitating the mechanism of action of other ingredients in the traditional formula.

Areca catechu (betel nut), another ingredient of the traditional formula, is commonly used in Asian medical traditions. Throughout Asia, betel chewing is believed to produce a sense of well-being, euphoria, warm sensation of the body, sweating, salivation, palpitation, heightened alertness and increased capacity to work. This collective experience across time would suggest that *Areca catechu* consumption affects the central and autonomic nervous systems.

Loss of an Important Antimalarial

The failure to consult classical Chinese medical sources and to consider the potential clinical validity of the combination of herbs in the original *changshan* mixture in favour of a determined search for an active ingredient in D. febrifuga resulted in the emetic properties of D. febrifuga leading to nausea and non-compliance in patients and the resultant abandonment of an important antimalarial drug. The significance of this loss has recently been underscored by a World

Health Organization (2002) regional publication which reports that a synthetic chemical derived from febrifugine (Df-1) is one of the two most powerful anti-malarial compounds of 4,700 screened. The other, derived from the traditional Chinese febrifuge, *Artemisia annua* (*qing hau*) is an endoperoxide synthesised from artemisinin (http://www.wpro.who.int/malaria/docs/shanghai.pdf).

At the heart of the *changshan* R&D exercise was a scientific debate that is still of central significance in the scientific evaluation of traditional medicines; it stands as an important theme in the medical and scientific history of drug development.

The contrast was between, on the one hand, a biomedical commitment to the identification of single active agents from plants, and, on the other, classical Chinese medicine's position that there are intended therapeutic synergies between the multiple constituents of a plant and among the many plants contained in traditional complex mixtures. Underlying the way in which the science on *changshan* has proceeded has been the consistent tension between tradition and modernity in medicine and the disparity in power relationships between modern medical exponents and the custodians of traditional medical knowledge. The low esteem in which Western-trained Chinese doctors held their traditional counterparts in Kuomintang China resulted in a power imbalance playing out in a research agenda (Lei 1999). To a significant extent this same dynamic continues internationally to the present, where indigenous explanatory models are seen as, at best, clues to the discovery of important active ingredients rather than as clinically viable and insightful expressions of experientially based polypharmacy.

Of central importance in any programme of scientific evaluation of traditional herbal medicines is an open-minded and respectful dialogue between modern medical scientists and the custodians of traditional medical knowledge. Such dialogue offers the potential to recognise within traditional explanatory frameworks new pathways and modes of action, potential interactive effects of the multiple ingredients in traditional complex herbal mixtures, and methods for capturing these effects in appropriately and sensitively designed research.

In the case of *changshan*, a vital dialogue that should have taken place to refine and focus the research questions that guide research and development appears not to have occurred. As the modern history of *changshan* illustrates, intrinsic power imbalances, compounded by mutual incomprehension, led to traditional Chinese explanatory models being overlooked as sources of valid research hypotheses, in favour of the more conventional assumptions made by modern medical investigators.

The result was not just one of a dominant scientific interest group prevailing over a weaker group. Rather, the resultant outcome was a significant scientific oversight, resulting in the loss of an important new, affordable and easily produced antimalarial treatment. The simple addition of traditionally used ingredients, such as ginger and liquorice, may have offset the emetic effects of *Dichroa febrifuga* or its purified alkaloids.

Clearly, *changshan* deserves to be reappraised in the light of its traditional formulation, including the anti-nausea agents and processes classically employed. The case of *changshan* also highlights the possible plethora of false negatives that may exist in the many plants rejected in contemporary antimalarial screening programmes. It raises the prospect that many other plants, and classical combinations of these, may merit renewed evaluation according to culturally and paradigmatically sensitive research methods.

Intellectual Property Rights

The interface between traditional knowledge, indigenous concepts of science and biomedical science is seen most prominently on the world stage in the arena of intellectual property rights pertaining to traditional medical knowledge and products. Recent years have seen heated international debate and legal challenge over the patenting of traditional knowledge (TK) and its products, such as grain species (see Chapter 9), traditional medicines, traditional art images, music and rituals. At the heart of this have been two conflicting forces (see Chapter 10): one has been the attempt by non-indigenous individuals and organisations to claim ownership of indigenous knowledge for commercial gain; the other has been the attempt of indigenous groups to fend off this trend and either to take ownership of such products themselves or to engage in partnership with fair sharing of benefits for the commercial development of their knowledge, products or processes (Dutfield 2001).

From the early 1990s to the present, this set of issues has played out in international legal forums, corporate board rooms, in political debate and, occasionally, in street demonstrations. At the heart of this conflict lies a perception that the patenting of indigenous knowledge represents an inversion of the intent of the patenting system. While the intent of patent systems is to provide protection to individual knowledge so that it can be brought out into the public domain for the public good, the patenting of indigenous knowledge may be seen as the reverse: taking knowledge that is already in the public domain and using legal means to place it into the private domain for the financial gain of a limited few.

These debates and challenges have taken place against the backdrop of two international legal frameworks which, until the present, have stood in apparent contradiction to one anther. These are the 1993 United Nations Convention on Biological Diversity (CBD), and the World Trade Organization's (2002) Trade Related Aspects of Intellectual Property Systems (TRIPS).

The CBD is the only major international convention that assigns ownership of biodiversity to indigenous communities and individuals and asserts their right to protect this knowledge. Article 8 (j) states that State Parties are required to

respect, preserve and maintain knowledge, innovations and practices of indigenous and local communities embodying traditional lifestyles relevant for the conservation and sustainable use of biological diversity and promote the wider application with the approval and involvement of the holders of such knowledge, innovations and practices and encourage the equitable sharing of the benefits arising from the utilisation of such knowledge, innovations and practices.

And Article 18.4 states that Contracting Parties should 'encourage and develop models of co-operation for the development and use of technologies, including traditional & indigenous technologies' (United Nations Convention on Biological Diversity, 1993).

The CBD competes for influence with the far more powerful TRIPS. TRIPS, now the key international agreement promoting the harmonisation of national IPR regimes, covers four types of intellectual property rights: (1) patents; (2) geographical indications; (3) undisclosed information (trade secrets) and (4) trademarks.

TRIPS makes no reference to the protection of traditional knowledge. It does not acknowledge or distinguish between indigenous, community-based knowledge and that of industry. TRIPS does not require adoption of UPOV (The International Union for the Protection of New Varieties of Plants) standards, but rather provision 'for the protection of plant varieties either by patents or by an effective *sui generis* system or by any combination thereof (Art. 27(3)(b). *Sui generis* systems are those that are developed according to the needs of a country or region. They are unique to that country and region.

Essentially, then, the CBD takes the view that if a product or process has existed in a culture for a long period of time, it is owned and hence protected under intellectual property law. By contrast, the view under TRIPS is that if it is not patented it is not owned. If it is not owned, it represents knowledge that is part of a global commons available for exploitation by all who so wish. (The CBD has published a summary of cases on access to genetic resources and benefit sharing at http://www.biodiv.org/programmes/socio-eco/benefit/case-studies.asp.)

The Doha Declaration and the Harmonisation of CBD and TRIPS

In November 2001 the declaration of the Fourth Ministerial Conference in Doha, Qatar, mandated a review of TRIPS provisions and called for a harmonisation between the CBD and TRIPS (www.wto.org). The Doha Declaration specifically requested the TRIPS Council to review the relationship between the TRIPS Agreement and the UN Convention on Biodiversity; the protection of traditional knowledge and folklore; and other relevant new developments that member governments raise in the review of the TRIPS Agreement. It added that the TRIPS Council's work on these topics is to be guided by the TRIPS Agreement's objectives (Article 7) and principles (Article 8), and must take development fully into account.

This process is still in progress at the time of writing. However, it is clear that researchers evaluating traditional medicines need to recognise that under international law, the customary owner, and often the country of origin, holds rights over the knowledge being evaluated (Bodeker 2003).

Conclusion

While the theoretical frameworks of traditional medicine and biomedicine differ radically in their views on the cause of disease and mechanisms of action of a particular therapy, it is apparent that the interface between these two perspectives on life and its course continues to be of central significance to the healthcare of a large sector of the world's population.

New insights from physics go beyond the current physiological and chemical explanatory models employed by mainstream biomedicine and may offer a common ground from which a unified view might be forged between biomedicine with its focus on the physical, and traditional medical systems, with their view that energy is the ground state of physical function, dysfunction and equilibrium.

Progress is unlikely to be linear, with a continuum of paradigmatic interfaces likely to continue for some decades to come – namely consultation and collaboration; isolated lab studies with no consideration of traditional views; traditionally grounded studies with a poor grasp of scientific principles or rigour; and commercially exploitative research and development with little or no consideration or customary ownership or entitlement.

In both the short and long term, culturally sensitive rapprochement between biomedical and traditional medical perspectives – their theories, their taxonomies, their approaches to research, and a mutually respectful dialogue in designing strategies for gathering evidence and sharing benefits that is meaningful to both sets of interests – promises much of value to medical theory, clinical medicine and public health.

References

Akimoto, M., M. Kimura, A. Sawano, H. Iwasaki, Y. Nakajima, S. Matano and M. Kasai 1986. 'Prevention of Cancer Chemotherapeutic Agent-induced Toxicity in Postoperative Breast Cancer Patients with Glycyrrhizin (SNMC)', *Gan No Rinsho* 32(8): 869–72.

Berlin, B. 1973. 'The Relation of Folk Systematics to Biological Classification and Nomenclature', *Annual Review of Ecology and Systematics* 4: 259–71.

Berlin, B. and E.A. Berlin 1996. *Medical Ethnobiology of the Highland Maya of Chiapas, Mexico: The Gastrointestinal Diseases*. Princeton, NJ: Princeton University Press.

Bodeker, G. (ed.) 1996. The GIFTS of Health Reports, *Journal of Alternative and Complementary Medicine* 2(3): 397–405 & 435–47.

———— 2001. 'Lessons on Integration from the Developing World's Experience', *British Medical Journal* 322: 164–67.

———— 2003. 'Traditional Medical Knowledge, Intellectual Property Rights and Benefit Sharing', *Cardozo Journal of International and Comparative Law* 11(2): 785–814.

Bodeker, G., C.-K. Ong, G. Burford, K. Shein and Y. Maehira (eds) 2005. *Global Atlas of Traditional, Complementary and Alternative Medicine*. Geneva: World Health Organization.

Chaudhury, R.R. 2002. 'A Clinical Protocol for the Study of Traditional Medicine and Human Immunodeficiency Virus-related Illness', *Journal of Alternative and Complementary Medicine* 7(5): 553–66.

Chen Fangshi 1944. 'The Preliminary Report on Clinical Research on Changshan for Treating Malaria (II)'. In *Changshan Zhinue Chubu Yanjiu Baogao* [The Preliminary Research Report on Changshan for Treating Malaria]. Nanjing: Guoyao Yanjiusi [Research Laboratory on National Drugs], Central Politics School.

Dutfield, G. 2001. 'Trade-related Aspects of Traditional Knowledge', *Case Western Reserve Journal of International Law* 33(2): 239–91.

Fairbairn, J.W. and T.C. Lou 1950. 'A Pharmacognostical Study of *Dichroa Febrifuga* Lour.: a Chinese Antimalarial Plant', *Journal of Pharmacy and Pharmacology* 2: 162–77.

Guoyao Yanjiushi 1944. *Changshan Zhinue Chubu Yanjiu Baogao* [The Preliminary Research Report on the Anti-malarial Drug Changshan]. Nanjing: Guoyao Yanjiushi [Research Laboratory on National Drugs], Central Politics School.

Hankey, A. 2001. 'Ayurvedic Physiology and Etiology: Ayurvedo Amritanam. The Doshas and their Functioning in Terms of Contemporary Biology and Physical Chemistry', *Journal of Alternative and Complementary Medicine* 7: 567–73.

———— 2004. 'Are We Close to a Theory of Energy Medicine?', *Journal of Alternative and Complementary Medicine* 10: 83–87.

Henderson, F.G., C.L. Rose, P.N. Harris and K.K. Chen 1948. 'g-Dichroine, the *Antimalarial Alkaloid of Chang shan*', *Journal of Pharmacology and Experimental Therapeutics*, 95.

Koepfli J.B. 1947. 'An Alkaloid with High Antimalarial Activity from *Dichroa febrifuga*', *Journal of the American Chemical Society* 69: 1837.

Lei, S.H. 1999. 'From Changshan to a New Antimalarial Drug', *Social Studies of Science* 29(3): 323–58.

Milliken, W. 1997. 'Malaria and Antimalarial Plants in Roraima, Brazil', *Tropical Doctor* 27 (suppl.1): 20–25.

Ni, M. (trans.) 1995. *The Yellow Emperor's Classic of Medicine. A New Translation of the Neijing Suwen with Commentary*. Boston, MA: Shambhala Press.

Pinzon, C. and G. Garay 1990. 'Por los senderos de la construccion de la verdad y la memoria', in *Por las Rutas de Nuestra America*, ed., C. Pinzon and G. Garay. Bogota, Colombia: Universidad Nacional de Bogota.

Sharma, H. and C. Clark 1998. *Contemporary Ayurveda: Medicine and Research in Maharishi Ayur-Veda.* Philadelphia: Churchill-Livingstone.

Tonkin, I M. and T.S. Work 1945. 'A New Antimalarial Drug', *Nature* 156: 630.

United Nations Convention on Biological Diversity (1993). http://www.biodiv.org

Unschuld, P. 1992. *Medicine in China: A History of Ideas.* University of California Press.

Venkatesananda, S. (trans.) 1984. *The Supreme Yoga: Yoga Vasishta.* South Fremantle, Western Australia: The Chiltern Yoga Trust (Aust.).

Warrell, D.A. 1997. 'Herbal Remedies for Malaria', *Tropical Doctor* 27 (Suppl 1): 5–6.

Williamson, E.M. 2001. 'Synergy and Other Interactions in Phytomedicines', *Phytomedicine* 8(5): 401–9.

World Health Organization Traditional Medicine Strategy. 2002–2005. May 2002. http://www.who.int/medicines/organization/trm/orgtrmmain.shtml.

World Health Organization Regional Office for the Western Pacific (WPRO). 2002 *Report: Meeting on Antimalarial Drug Development, Shanghai, China, 16–17 November 2001, (WP)MVP/ICV/MVP/1.4/001.E.* Manila, Philippines: World Health Organization Regional Office for the Western Pacific. http://www.wpro.who.int/malaria/docs/shanghai.pdf

WHO (World Health Organization). 1978. *Traditional Medicine.* Geneva: World Health Organization.

World Trade Organization. 2002. *Trade Related Aspects of Intellectual Property Systems* (TRIPS), at www.wto.org/english/tratop_e/trips_e.htm

Zhang, X. 2000. 'Integration of Traditional and Complementary Medicine into National Health Care Systems', *Journal of Manipulative and Physiological Therapeutics* 23(2): 139–40.f

3 Local and Scientific Understanding of Forest Diversity on Seram, Eastern Indonesia

Roy Ellen

Introduction

Foresters, biogeographers and tropical forest ecologists have devised increasingly sophisticated classifications of forest types (e.g. Eyre 1980). Forest 'types' and their more localised and discrete components, which might variously be described as 'habitats', 'niches', 'biotopes' and 'ecotones', constitute what ecologists understand by 'secondary biodiversity': that is diversity in terms of associations of species rather than the ('primary') diversity measured in terms of the numbers of species (or other taxanomic categories). Although the classifications of foresters in particular have been largely determined by the practical considerations of the industry, during the latter part of the twentieth century they have been much influenced by the developing science of forest ecology, and the technologies of remote sensing (Howard 1991) and Global Information Systems (GIS) (C.A. Johnston 1998). The typologies of forest ecologists, while originally rooted in those of foresters, have become increasingly distanced from them in an attempt to model more accurately the dynamic character of forest diversity.

Although the pragmatic schemes used by national forest departments have often responded to local situations by incorporating categories which anthropologists would describe as 'folk', 'emic' or 'indigenous', on the whole the practice of modern forestry has markedly diverged from the representations of secondary biodiversity which these imply (see e.g. Muraille 2000: 74–77). I shall show in this chapter how the categories and coordinates applied by Nuaulu in eastern Indonesia contrast with most official functional classifications of tropical forest type in being dynamic, multidimensional, not tied to complex nomenclatures, and unashamedly *local*. I will also show how they anticipate recent modelling attempts in scientific ecology which emphasise the 'patchiness' of tropical rainforest.[1] Studies of tropical forest peoples have revealed not only an extensive native knowledge of trees, but also local recognition of forest diversity and the existence of coherent vernacular classifications of forest types. The evidence suggests some variation in nomenclatural and classificatory patterns. Work in the Amazon region, for example, has reported folk classifications of considerable complexity (e.g. Fleck and Harder 2000: 1–3; Shepard et al. 2001; Shepard et al. 2004), which do not appear to be matched by comparable data from, for example, southeast Asia and New Guinea (Sillitoe 1998; Ellen, unpubl.). However, my concern here is with the commonalities which such studies yield, using Nuaulu data as a point of departure; and with a comparison between ethnoecological classifications of tropical forest in general and those offered by scientists and officials. Thus, we are dealing with issues of *scale*, which as Sillitoe (2002a, 2002b) has shown, have increasingly become critical in judging the appropriateness of local and global knowledge respectively in the context of development practices.

There is one further important point which needs to be registered before embarking on this specific analysis, which is that technical forestry practice, especially including the typologies it has devised, was first formalised and institutionalised, and indeed continues, within an overt political context, which has shaped its underlying assumptions. Thus, in Indonesia, as in many other places, the definition and demarcation of land as 'forest' during the colonial period can be seen as a very concrete 'territorialisation strategy' in which first the colonial state (Boomgard 1994) and then an independent republic manifested its existence and legitimated its jurisdiction. Through the 'adminstrative ordering of nature' the remit of the state was made 'legible' (Scott 1998: 2, 4). Thus, forestry departments became one of the most important agencies in territorialising state power, and inevitably in doing so had the effect of simplifying the 'illegible' cacophony of local property regulations and communal tenure, which presented itself as an administrative nightmare (Scott 1998: 35, 37). But the process of simplification and of inscribing legibility had the consequence of excluding and including people within particular geographic boundaries and of controlling their access to natural resources; it provided the inevitable grounds for conflicts with local perceptions and values. In such a situation 'tribal' peoples, already a 'problem' because of their administrative peripherality, became additionally so through a forestry policy defined in such a way as to deny any merit in forms of extraction other than for timber, and which especially demonised swiddening or long fallow forms of agriculture (Dove 1983, Dove et al., this volume; Persoon et al. 2004: 26). And whereby the forested territory of the modern state could be understood in basically linear, abstract and homogenous terms, as the monothetic management of wood, this is quite the opposite of how local people experience forest, or indeed any other kind of space (Vandergeest and Peluso 1995: 388–89; c.f. Sivaramakrishnan 2000).

The main evidence which legitimated this new control, and therefore the problems that subsequently arose, were maps (c.f. Dove et al., this volume). In other words, the 'territorialisation strategy' only effectively became a reality as competent cartographic surveys were conducted, especially, but not exclusively, those which remapped forest on the basis of scientific criteria, such as soil type, slope, vegetation and timber utility, or what Vandergeest and Peluso (1995: 408) call 'functional territorialization'. In the context of Indonesia, this emergence of an ideology of state forestry and of state and scientific classifications of land, as well as the very idea of the 'management' of natural resources (Ellen 2003), happened first in colonial Java, with the establishment in 1808 of the 'Dienst van het [Ost-Indische] Boschwezen' and the 'Administratie der Houtbosschen' (Anon. 1917: 390; Departemen Kehutanan 1986; Peluso 1992: 6–8, 44–45; Boomgard 1994: 119). In the distant Moluccas, and on Seram in particular, such practices did not become a reality until the Topografische Dienst survey of 1917, which I shall return to in the final section of this chapter. But the conceptions of forest space which accompany this strategy are differentiated from pre-existing local conceptions, again in terms of scale.

Methods for Studying Local Representations and Understandings of Forest Diversity

In order to understand how Nuaulu conceive and use their classification of types of forest it is important to show in detail the composition and ecological character of the kinds of forest which they label. This entails the use of a plot methodology, in which all flora above a certain size and other features occurring in a specified area are logged, mapped and named with the help of local people. But one of the problems of comparing different compositional studies of tropical forest on a global scale has been inconsistency in the size of the plots, or quadrats, employed. The majority of ecological studies have relied upon plots of between 0.63 and 1.2 ha, most commonly 1 ha (e.g. Valencia et al. 1994; Richards 1996). While the problems of quadrat surveys generally have been widely discussed (e.g. Kershaw 1973; Kent and Coker 1992), in ethnobotany the problems are, if anything, greater, despite there being a smaller number of studies to which to refer (Martin 1995: 157–59). Rectangular plots, where one side is considerably longer than the other, have been used by Boom (1989), for example belt transects of 10 m by 1,000 m in his work amongst the Chácobo of northern Brazil, and 40 m by 10 m plots have been used by Puri (2005), working among Penan Benalui in east Kalimantan. But most plots have conventionally been square, for example the use by Salick (1989: 191) of stratified random 5 m by 5 m plots to sample Amuesha swiddens. Bernstein et al. (1997) used 0.23 ha (48 m by 48 m) plots divided into four quadrats in their work amongst the Brunei Dusun. Allan (2002: 137), in her Guyana work, used a plot size of 0.25

ha (50 m by 50 m) because the locally defined (largely Makushi) forest types often did not extend over areas large enough or symmetrical enough to allow a larger plot to be established within the forest type. Sillitoe (1998) has even used 10 m by 10 m quadrats to measure differences in tree flora.

Although the size of a plot must ultimately be determined by research objectives and practicalities, Greig-Smith (1964: 28–29) identifies two problems with smaller quadrats: that there is a greater chance of significant edge effects (due to an observer consistently including individuals which ought to be excluded); and that the frequency distribution for individuals is more likely to be Poisson than normal, with the magnitude of the variance related to the mean. This latter makes it difficult to apply some of the usual statistical procedures for comparing populations. The first effect can be corrected by including individual trees which fall on the edge of the plot only if 50 percent or more of their canopy area is judged to fall within the plot; otherwise they are excluded. But although fieldworkers may make every effort to apply this rule consistently, it is inevitably a subjective assessment, and there will always be an unknown level of observer error. This must be acknowledged as a limitation of the data collected in plots of this size. This second problem identified by Greig-Smith can be tested for and potentially corrected using data transformation.

In 1996 the Nuaulu were a group of some 2,000 individuals, engaged in swidden cultivation, sago extraction, hunting and forest extraction in lowland central Seram in the Indonesian province of Maluku, the Moluccas (Figure 3.1). In that year I conducted eleven plot surveys in forest which Nuaulu exploited and which was acknowledged as belonging to them. The intention was to sample from as wide a range of mixed forest vegetation as possible with which Nuaulu were actively interacting. Deliberately excluded from the sample area were mangrove, littoral biotopes, groves and plantations, recently abandoned garden land, and swamp forest predominantly covered in sago (*Metroxylon sagu*). Also, because Nuaulu seldom extract from forest above 1,000 metres above sea level, mountain habitats above this altitude were excluded. As in some earlier studies (Allan 2002, 1995; M. Johnston 1998; Sillitoe 1998), the object of the surveys was to obtain botanical composition data for locally recognised forest types, and for this reason plots were placed within areas identified by local informants as indicative of a particularly salient forest type, and within the range of that type, in locations which were relatively accessible. This inevitably resulted in a non-random, nonsystematic distribution, which limited the quantitative analysis that could be performed on the data, and reduced their value as sources for a general ecological survey of the forest. It must, therefore, be borne in mind that the aim was an analysis of ethnoecological knowledge of different emically defined areas and not a study of forest ecology. The general characteristics of the plots surveyed and their geographic location are indicated in Table 3.1 and Figure 3.2 respectively.

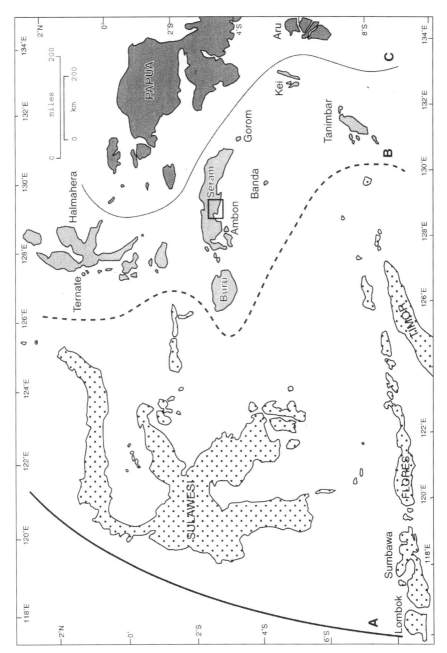

Figure 3.1 Seram in the context of the eastern Indonesian archipelago, showing places mentioned in the text. Line A represents Wallace's Line of Faunal balance. Line B is Weber's Line, and Line C is the western boundary of the Australo-Pacific region. The box indicates the area enlarged in Figure 3.2.

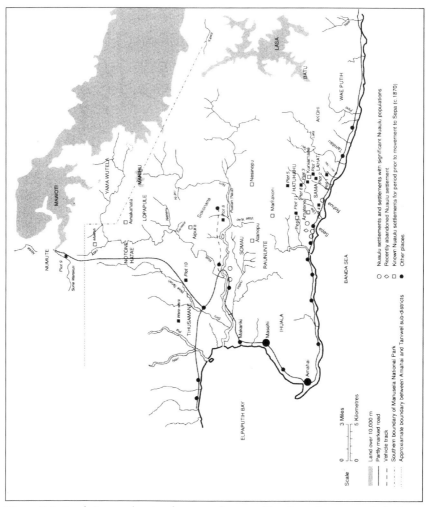

Figure 3.2 South Seram, showing former and current Nuaulu settlements and plot locations mentioned in the text, as of 1996.

Table 3.1 Basic data on location and description of plots.

Plot number	Toponym	Location	Nuaulu description	Summary ecology	Altitude (m)	Distance from Rohua (km)
1	Nahati sanai	Near river Makoihiru	*nisi ahue*	Edge of cultivated area, depleted mixed forest	200	3.75
2	Mon	On river Mon, near Rohua	*wesie*	Edge of cultivated area, depleted mixed forest	150	4.25
3	Nahane hukune	Sama, north of Rohua	*wesie*	Depleted, near logging road	200	2.00
4	Ratipanisa	On river Yoko, near Rohua	*nisi ahue*	20-year-old regrowth garden land	100	0.75
5	Besi	Between Iana Ikine and Iana Onate	*nia monai*	Old settlement site	300	3.00
6	Iana onate matai	near river Upa	*nisi ahue*	30-year-old regrowth garden site	50	2.50
7	Mon sanae	Head of river Mon, near Rohua	*sin wesie*	Protected sacred forest (clan Peinisa)	300	4.50
8	Sokonana	Near transmigration site on river Ruatan	*wesie*	debris of cut trees, far from gardens near new road	100	25.0
9	Rohnesi	W of trans-Seram highway near Wae Sune Maraputi	*wesie*	High-altitude undisturbed forest	900	70.0
10	Wae Pia	E of trans-Seram highway near Tihasamane, Tanaa valley	*wesie*	Lowland forest	200	55.0
11	Amatene	Above headwaters of Iana river, trib. of Upa	*nia monai*	Old village site	400	5.00

a study of forest ecology. The general characteristics of the plots surveyed and their geographic location are indicated in Table 3.1 and Figure 3.2 respectively.

Plots 1 to 6 were each 400 m^2 (20 × 20 m), and plots 7, 8, 9 and 11 were each 900 m^2 (30 × 30 m). Plot 10 was 430 m^2. The four large 900 m^2 plots were surveyed because particular features of the plot were judged to be intrinsically interesting: plot 7 being sacred protected forest, plot 8 an area on the fringe of recent settlement, plot 9 a high-altitude site traditionally used for collecting *Agathis* resin, and plot 11 an old village site. All measurements were of surface areas, but surfaces which were often on steep slopes. Although angle of slope was measured, plot maps (e.g. those in Ellen, unpubl.) and density data were accordingly distorted: the steeper the slope the greater the distortion. Another problem associated with plot size was that small plots tended to underestimate species richness compared with larger plots in the same area. Figure 3.3 shows the relationship between the number of species and size of plot, from 0.05 to 0.5 ha, as used in a number of Moluccan studies. The inclination of the species–area curve is roughly consistent with species-area curves obtained in other studies of lowland rainforest in island southeast Asia, though Edwards et al. (1990: 168, fig. 15.2 (a)) found that curves in the Manusela National Park generally flattened out at 0.25 ha, suggesting that enlarging plot size further would not have added more species. I shall return to a consideration of the implications of this pattern in the next section.

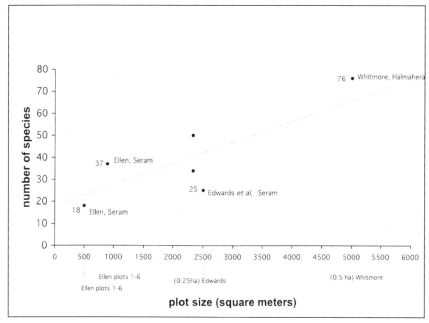

Figure 3.3 Species numbers in relation to plot size for various forest composition studies in the Moluccas.

Each plot was surveyed with a minimum team of three adult males. They were not always the same persons, but there was a marked overlap in membership. All team members were trained in the use of measuring, marking and enumeration techniques before each survey. Plots identified by me were first measured and marked up using a 30 m fibreglass retractable tape and spray paint. Plants were included in the survey if they were 10 cm dbh (diameter at breast height) or above, effectively restricting the census to trees and large lianas. The 10 cm threshold has become standard practice, although it has been demonstrated (Valencia et al. 1994) that only counting trees over 10 cm dbh can underestimate the diversity of woody species present as saplings. The location of each tree above 10 cm dbh was marked on prepared graph paper and an independent Nuaulu identification sought from each field assistant present. If there was any disagreement, discussion was allowed to see whether agreement might be reached or whether informants would agree to disagree. Where possible, voucher specimens were collected, including bark, but seldom (for tall trees) fertile specimens. We would then move on to the next tree, and the same sequence would be repeated. Back in the village, voucher specimens would be further discussed, fully documented, preserved by drying or in alcohol-soaked newspaper. All specimens, collected in triplicate where possible, were checked and sorted at the Herbarium Bogoriense. One set was retained in Bogor, a second was sent to Kew, and the third set retained in the Ethnobiology Laboratory of the University of Kent. All systematic data were entered into the Nuaulu Ethnobotanical Database (NED). In the field, local names obtained during plot surveys were checked against earlier data entered into the NED, and revisions made as necessary, often involving further consultation with informants. Where voucher specimens were absent or insufficient, photographs, drawings and visual descriptions were used in combination with standard reference manuals. Where Nuaulu were also able to provide Ambonese Malay terms, these were matched where possible against standard lists of Moluccan tree species with vernacular glosses (e.g. Whitmore et al. 1989), but always back-translated several times in different contexts to minimise erroneous determinations. Table 3.2 lists the total number of standing trees recorded for each plot compared with: (a) numbers of trees for which Nuaulu informants could provide names, and (b) botanical identifications to different levels of taxonomic specificity obtained from the various authorities consulted. Although phylogenetic identifications have not been obtained for all vouchers (in some cases even to generic level), and there are some plot trees for which vouchers were not obtained, the general pattern of identifications demonstrates a strong correlation between vernacular names provided and scientific species, suggesting that measures of species density, for example, might reasonably rely on vernacular names as proxies where determinations are unavailable.

Table 3.2 Summary of selected 1996 plot data: levels of identification.

Plot	1	2	3	4	5	6	7	8	9	10	11
1	N	34	30	37	26	28	54	124	116	80	35	68
2	N with Nuaulu name[a]	34[b]	28	37	26	28	54	124[b]	115	74	35	68[b]
3	Percentage N identified with agreed Nuaulu name	100	93	100	100	100	100	100	99	94	100	100
4	N Identified to family level	30	25	35	25	16	50	108	98	54	25	68
5	N identified to generic level	20	21	32	24	16	49	103	90	37	25	58
6	N identified to species level	5	12	19	24	1	38	46	41	29	12	40

Key: N= number of standing trees 10 dbh or above; [a] = name agreed by minimum of three adult male field assistants; [b] = one doubtful

The Ecology of Lowland Rainforest on Seram and the Classification of Vegetation Types

If we compare the forest composition of Seram with that of the large islands of western Indonesia (Borneo and Sumatra) and New Guinea, it is clear that Seram lies in a zone of transition (Wallacea) between the predominantly Dipterocarp forests of Sunda (Asia) and the Australo-Pacific tree flora of Sahul (Oceania) (Figure 3.1). As we move eastwards Dipterocarpaceae fade out and are replaced by other characteristic families, such as (in lowland areas) Myrtaceae (particularly the distinctive Eucalypts), Myristicaceae, Lauraceae and Guttiferae (Glatzel 1992: 17–18; Edwards et al. 1993: 66, table 2a). This same pattern is confirmed by my own data (Table 3.3), with the most numerous families represented in the Nuaulu plots being Myrtaceae, Myristicaceae and Guttiferae.

The Manusela National Park study, conducted in 1987 (Edwards et al. 1993), was primarily concerned with altitudinal variation. It was based on nine 0.25 ha (i.e. 50 m by 50 m) permanent sample plots at a range of altitudes from sea level to 2,500 m asl (above sea level) south of Gunung Binaia on the central mountain spine. There were two sequences, one over calcareous rocks and the other over non-calcareous rocks. Both show that soil pH decreases with altitude while organic carbon increases. In addition to a lowland zone (the primary focus of the present analysis) the Manusela study yielded data distinguishing alpine, subalpine (characterised by shrubby *Rhododendron*), montane (high-altitude tree fern grassland), and lower montane zones, the latter with an upper band dominated by Myrtaceae and Lauraceae, and a lower band by Fagaceae. The dominant species of lowland and lower montane forest (which occurs at a lower altitude on smaller mountains, and is therefore of some interest to us here) are reported for four plots as, respectively: *Drypetes longifolia, Planchonella nitida* and *Astronia macrophylla; Lithocarpus* sp., *Litsea robusta* and *Shorea* sp.; *Lithocarpus* sp. and *Weinmannia*; and *Phyllocladus hypophyllus*, Myrtaceae and *Trimenia*. By comparison, in the eleven Nuaulu plots (Table 3.3, Table 3.4), there are 39 families represented overall. There are no clear dominants in one, and in the others the dominants are Euphorbs and *Syzygium; Shorea selanica; Artocarpus integer*; Myrtaceae and *Mallotus; Areca catechu*; Myrtaceae and *Annona reticulata, Polymatodes nigrescens* and *Myristica; Calophyllum inophyllum; Syzygium* and *Myristica*; and finally Myrtaceae and *Macaranga involucrata*. In other words, despite a strong similarity between the Manusela data and my own at the family level, the only overlap of dominants at the generic level is with respect to *Shorea* (a genus which is numerically quite untypical of the Moluccas) and the important Myrtaceae genera, no doubt including *Syzygium* and *Eugenia*. The explanation for this difference may in part lie in the deliberate bias in the Nuaulu plots in favour of anthropogenic vegetation, but it also reflects the general diversity and patchiness of species composition of lowland rainforest on Seram below 1,000 m (mostly on low hill land), which had been evident from my work in 1970–71, at a time

Table 3.3 Family frequencies for individual trees in 1996 plots.

Family	Plot 1	Plot 2	Plot 3	Plot 4	Plot 5	Plot 6	Plot 7	Plot 8	Plot 9	Plot 10	Plot 11	Total
MYRTACEAE	6	1	4	1	6	1	16	17	11	8	15	86
MYRISTICACEAE		2	2		2	7	6	21		6		46
EUPHORBIACEAE	13	2		1	5	2	7		3	1	11	46
GUTTIFERAE	1		5				5	8	17	4		40
PALMAE		4				21	4	1	2	1	3	35
MORACEAE	1			17	1		8			2	3	33
LEGUMINOSAE	1	1	1	1		3	4	3	4		7	25
POLYPODIACEAE								23				23
LAURACEAE	1	2	1	1			7	2	4			18
BURSERACEAE			4	2		1	7	3			1	18
ANNONACEAE		2				1	11	3				17
EBENACEAE	2		1		1	1	8	1				14
LEEACEAE		3			1	3	5					12
RUBIACEAE	3	2				1	4		1			11
ROSACEAE			3			1	2	1	3			10
DIPTEROCARPACEAE			10									10
LOGANIACEAE						1			7			8
OLEACEAE			2					6				8
VERBENACEAE	2			1			2		2	1		8
STERCULIACEAE						3	1	1			2	7
MELIACEAE		3	1		1		2					7
LECYTHIDACEAE			1					5				6
APOCYNACEAE							2			3		5
PANDANACEAE						2					3	5
ARAUCARIACEAE									4			4
MELASTOMACEAE							1	2		1		4
GRAMINAE							1				3	4
ELAEOCARPACEAE			1								3	4
MALVACEAE											3	3
ANACARDIACEAE							1				2	3
FLACOURTIACEAE						3						3
Eight families with two or less trees[3]	1	1	1	1	3	1	3	1		13		
Unidentified	4	4	5	2	1	12	4	16	18	27	10	108

when there were no detailed studies of forest ecology for Seram (Ellen 1978: 67–68). Table 3.4 provides a summary of 1996 Nuaulu plot data for floristics and forest structure (see also Figures 3.4, 3.5 and 3.7).

Both montane and lowland plots surveyed on Seram display low species diversity compared with many rainforests in the far east (Whitmore 1984), species declining with altitude above 600 m. If we take the mean of all plots (n = 4) within the altitude range 0–1900 reported by Edwards et al. (1993) – that is, covering approximately the same altitudinal range as the Nuaulu plots, and including both lowland and sub-montane areas – the number of species per plot is 25. Species number for the Nuaulu plots was higher, even though plot size was smaller. Assuming an approximately proportionate increase in species number with plot size (Figure 3.3), Nuaulu species numbers for a 0.25 ha plot would likely be between 35 and 50, and projected to 0.5 ha then around 75. In terms of the index of species richness (d = S/√N), the Manusela data give a mean of 2.14 (range = 1.44 > 2.99, where N = 4) and the Nuaulu data a mean of 3.32 (range = 1.7 > 5.6, where N = 11). Whitmore et al. (1987), in an enumeration study carried out on Halmahera at 630 m asl, recorded 76 species > 10 cm dbh, from 31 families within a 0.5 ha plot, giving a species richness index of 3.94, slightly higher than both the Manusela and Nuaulu data. Sidiyasa and Tantra

Figure 3.4 Looking eastwards along the Nua valley towards Mount Binaiya from Notone Hatae on the Trans-Seram Highway; midway between the south coast and Sawai on the north coast, but on the southern watershed. Apart from the roadside strip, all forest here is described simply as *wesie*, and consists of long-term regenerated forest and forest which has not obviosuly been modified by humans. March 1996 (96-11-20).

Table 3.4 Summary of plot data: floristics and forest structure.

Plot	Size (m²)	Alt.	N	S	N/m²	S /m²	H	E	1/D	Most frequent family/genus (N reported for taxon)
1. Nahati	400	200	34	20	0.085	0.05	2.75	0.92	18.10	EUPHORBIACEAE (13)
2. Mon	400	150	30	21	0.075	0.053	3.01	0.99	48.33	MYRTACEAE, *Syzygium sp* (6) PALMAE (4)
3. Nahane hukune	400	200	37	16	0.093	0.04	2.47	0.89	10.74	DIPTEROCARPACEAE *Shorea selanica* (10) GUTTIFERAE *Calophyllum inophyllum* (5)
4. Yoko	400	100	26	9	0.065	0.023	1.41	0.64	2.39	MORACEAE *Artocarpus integer* (17)
5. Benteng	400	300	28	9	0.07	0.023	1.77	0.80	4.97	MYRTACEAE (6) EUPHORBIACEAE *Mallotus sp.* (5)
6. Yana Onate	400	50	54	22	0.14	0.055	2.79	0.90	13.25	PALMAE (mainly *Areca catechu*) (21) MYRISTICACEAE (7)
7. Mon sanae	900	300	124	47	0.14	0.052	3.62	0.94	38.32	MYRTACEAE (16) ANNONACEAE *Annona reticulata* (11)
8. Sokonana	900	100	116	33	0.13	0.037	3.05	0.87	15.02	POLYPODIACEAE *Polymatodes nigrescens* (23) MYRTACEAE *Myristica* (17)
9. Rohnesi	900	900	80	28	0.089	0.031	2.92	0.88	14.77	GUTTIFERAE *Calophyllum inophyllum* (17)
10. Wae Pia	430	200	35	21	0.081	ʹ0.05	2.88	0.94	23.8	MYRTACEAE *Syzygium sp.* (8) MYRISTACEAE (Mainly *Myristica sp.*) (6)
11. Amatene	900	400	68	26	0.076	0.03	2.98	0.91	18.67	MYRTACEAE (15) EUPHORBIACEAE *Macaranga involucrata* (11)

Key: N = number of trees, S = number of species (see *Note*) = species richness, N/m² = tree density, S/m² = species density. H = Shannon diversity index $\Sigma Pi(\ln Pi)$, where Pi = the proportion of individuals in the ith species. The value of H increases with both species richness and equitability (evenness with which individuals are distributed among the species). E = equitability H/lnS, 1 being where individuals are most evenly distributed amongst the species, and 0 being the least. D = Simpson's index of dominance, $\Sigma(ni(ni-1) / N(N-1))$ (gives probability of any two individuals drawn at random from an infinitely large community belonging to different species, where ni = the number of individuals in the ith species. As D increases, diversity decreases. Simpson's index is therefore expressed as 1/D.

Note. S and formulae incorporating S are derived from Nuaulu vernacular terms for trees, and therefore an assumed equivalence between Nuaulu terms and taxonomic species.

Figure 3.5 Plot 8, riparian forest at Sokonana, north of the Ruatan river. February 1996 (96-08-25).

(1984), working in the northern part of the Manusela National Park (Wae Mual), provide a very low species richness index of 0.97 for lowland rainforest, even for Seram. In contrast, a 1 ha plot in Brunei (Poulsen et al. 1996) had 550 trees, 231 species and an index of species richness of 9.85. The low species richness values reported for Seram, compared with Borneo (see also Proctor et al. 1983) or New Guinea (Paijmans 1970), appear to be related to the recent geological emergence of the island and the varying levels of isolation over a six million year period (Audley-Charles 1993).

There have been various attempts to distinguish distinct forest types for Seram, beginning with the colonial forestry department, and its successor in independent Indonesia (Departemen Kehutanan). Figure 3.6, for example, shows superimposed official forestry categories on a map of South Seram. More instructive, from an ecological standpoint, is the typology presented by Glatzel (1992: 17–18), based on work conducted in West Seram by the Agricultural Faculty of Pattimura University, in the same area in which the ethnobotanical work of Suharno (1997) was subsequently conducted (Table 3.5). What is significant about these official typologies, as we shall see, is their general lack of congruence with local folk classifications.

We now know that tropical rainforest is a less stable and more diverse vegetational regime than once thought, and that a great deal of forest, especially lowland forest, is relatively recent regrowth, much of it following human interventions. Little was known about the time-scale of secondary successions at the time of Richard's classic benchmark study (1996: 400), and it was generally accepted that *primary* forest could be taken as mature old forest which had reached a fairly stable equilibrium or ecological succession (Spencer 1966: 39). The certainties of these older equilibrium and functional models, and the static pristine rainforest concept are no longer accepted in their entirety, and a single forest ecosystem type concept based on notions of a stereotypical or 'essentialised' climax forest are inappropriate (Johns 1990: 144; see also Blumler 1996: 31). Instead, contemporary models of lowland forest ecology emphasise more the patchiness, historicity and diversity of composition. Moreover, measures of what is understood by diversity have become more sophisticated. For example, it is now usual to distinguish alpha diversity (the number of locally occurring species) from beta diversity (diversity at the level of species communities). Some argue that alpha and beta diversity are related, alpha diversity resulting from a mosaic of juxtaposed niches and microhabitats. Consequently, to attempt to measure empirically the number of different 'types' of vegetation in a tropical rainforest may seem so time-consuming and ultimately subjective as to be hardly worth the effort (Condit 1996). This makes establishing simple typologies difficult, though there may be good practical reasons (in connection with forest management), to devise and recommend them. Thus, the authoritative classification of Pires and Prance (1985: 112–13), which draws extensively on ethnically diverse local ethnoecologies, distinguishes about twenty-two separate vegetation types for the

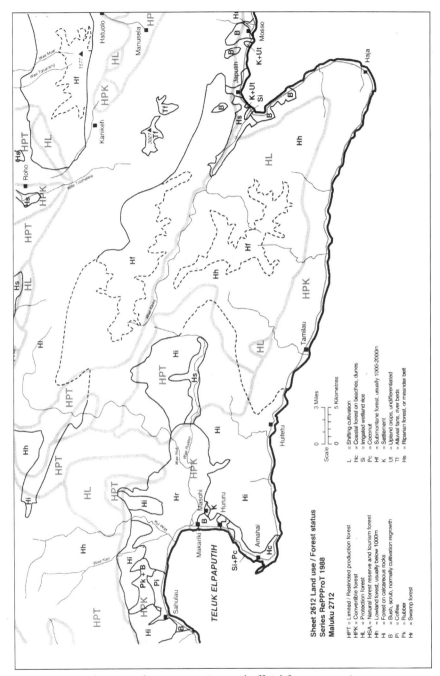

Figure 3.6 South Seram, showing superimposed official forest categories.

Table 3.5 Forest types distinguished by Pattimura University Agricultural Faculty survey team and utilised by Glatzel (1992)

	Species association	Notes
1.	*Imperata – Melaleuca* grassland	
2.	*Melaleuca – Imperata* woodland	As 1, but *Pteridium* also charactersistic
3.	*Strombosia phillipensis* dense evergreen forest	*Anthocephalus macrophyllus, Vitex cofassus, Octomeles sumatrana, Macaranga* sp. *Pandanus* on disturbed sites
4.	*Vitex – Pterocarpus* dense evergreen forest	No pronounced dominants. Most common species: *Vitex cofassus, Myristica insipia, Pterocarpus indicus, Eleaocarpus sphaericus.*
5.	*Anthocephalus – Intsia* dense evergreen forest	
6.	*Intsia – Pterocarpus* dense evergreen forest	
7.	*Canarium – Myristica* dense evergreen forest	Easy to walk in: dominants include *Canarium indicum, Myristica insipida and Pterocarpus indicus*
8.	*Puteria – Metrosideros* dense evergreen forest	No real dominants, but common species include *Puteria obovata, Metrosideros vera, Calophyllum inophyllum, Litsea* sp., *Elaeocarpus sphaericus, Cananga odorata, Dysoxilum cautostachyum and Octomeles sumatrana.*
9.	*Metrosideros vera* dense evergreen forest	Almost exclusively monospecific
10.	*Octomeles – Arenga* evergreen forest	Dominants are *Octomeles sumatrana, Ficus* sp., *Pinanga* sp., *Nauclea orientalis* and *Bambusa* sp. (*Arenga* very characteristic)
11.	*Metroxylon – Bambusa* evergreen forest	*Metroxylon, Bambusa* and *Arenga*
12.	*Sonneratia – Bruguira* evergreen forest.	Coastal and estuarine mangrove

Brazilian Amazon. However, it appears to make increasing sense to interpret forest composition in terms of the distinguishable kinds of process which lead to variation (Sprugel 1991; Fairhead and Leach 1998: 186). Rainforest, it is now widely acknowledged, is a mosaic of patches of different sizes, whether looked at in terms of different silvigenic stages of development (Torquebiau 1987), or differences based on substrate, altitude and aspect, soil water content, or dynamics arising from species dispersion.

Human Use and Modification of Forest

Much tropical lowland rainforest – in Indonesia as elsewhere – is now seen as the product of many generations of selective human interaction and modification (deliberate and inadvertent), optimising its usefulness and enhancing local diversity. The outcome is a coevolutionary process in which human activity is essential. Indeed, particular patterns of forest extraction and modification are often seen as integral to its sustainable future. For some authorities, the evidence for intentional rather than serendipitous human influence is so compelling as to invite the description of 'managed' forest (Clay 1988; Schmink et al. 1992: 7–8).

The empirical work supporting these claims comes mainly from the Amazon (e.g. Balée 1989, 1992, 1994; Anderson and Posey 1989); but more recently also from Africa (Fairhead and Leach 1996), and increasingly from large parts of Malaysia and the Western Indonesian archipelago (Rambo 1979; Dove 1983; Maloney 1993; Padoch and Peluso 1996; Aumeeruddy and Bakels 1994; Brookfield et al. 1996; Padoch and Peters 1993; Peluso 1996; Colfer et al. 1997; Puri 2005). On Seram, the generally low tree diversity is much influenced by disturbance, and there is abundant evidence that human agency has had consequences for forest ecology. This has been largely through the long-term direct impact of small-scale long forest-fallow swiddening, the extraction of palm sago (mainly *Metroxylon sagu*), and arboriculture over many hundreds of years, featuring a small number of crucial nut- and seed-yielding trees, most notably *Canarium, Aleurites, Pandanus* and *Celtis* (Ellen 1988, 2001). For example, the mature mixed forests of central Seram contain a higher proportion of *Canarium* than would be expected without human encouragement (Edwards 1994; Latinis 2000), and the high proportion of *Canarium* to other genera in the Nuaulu plot 7 (a protected area) is worth noting in this respect (Figure 3.8). The selective felling of large trees allows for small patches of characteristically secondary forest species (e.g. *Trevesia sundaica, Macaranga hispida, Artocarpus elasticus* and *Bombax* spp.), while species which are often regarded as being characteristically 'secondary' – *Prunea arborea, Platea excelsa* and *Chisocheton sandoricarpus* – have become characteristic of primary rainforest (Edwards et al. 1990: 171). In addition, the introduction through human agency of the pig (and almost certainly the cassowary), and in more recent centuries, deer too, has had a marked impact on forest dynamics, both in terms of the feeding patterns of these megafauna, and also through systematic human predation. Extraction of a wide range of useful products, including timber for local use, selective logging and collection for exchange (resins, rattan, birds) has also been significant (Ellen 1985: 563; Ellen 1999: 137). In addition to *Canarium*, among the endemic tree species whose distribution have been significantly affected by their importance in exchange are *Agathis dammara* (for its resin) at sub-montane altitude and *Melaleuca leucodendron* (for medicinal oil) in drier more open areas.

In earlier work (Ellen 1978: 67, map 8, p. 117) I have described the distribution of mature regrowth and 'primary' forest for 1970–71 in the vicinity of the Nuaulu settlement of Rohua. The distribution showed a striking visual distinction between (a), the bulk of forest stretching from a very minimum of around 100 m from the coast northwards and mountainwards, and (b) isolated patches apart from the major block of forest and forming barriers between cultivated areas. This latter residual distribution tends to be along ridges and around steep

Figure 3.7 Upland forest with *Agathis dammara*, near plot 9 (Rohnesi), west of Trans-Seram Highway near Wae Sune Maraputi. March 1996 (96-11-21b).

Figure 3.8 *Canarium commune* below the old village site of Amatene (plot 11). March 1996 (96-12-21).

knolls, unsuitable for cultivation under normal circumstances. As a result, these areas (together with the margins of the main forest block) within easy access of the village, are an important source of construction timbers and certain other products, leading to gradual thinning and denudation. Different kinds and degrees of extraction lead to different kinds of secondary regrowth, some directed (that is managed) and some arising by default (c.f. Sillitoe 1996: 216–24). These include: (1) young secondary growth: recently deserted clearings with rapidly growing herbs, shrubs and small trees of a relatively few number of genera – for example, *Trema orientalis, Euphorbia hirta, Homalanthus populifolius* and many kinds of pteridophyte; (2) medium secondary regrowth: one to ten years, with small trees gradually becoming dominant, for example, *Aleurites moluccana* and *Melastoma malabathricum*; (3) mature secondary forest, with a great variety of small and medium-sized trees, shrubs and vines in areas with over ten years of secondary growth; and finally (4) bamboo thicket, also found in combination with the three above associations.

The tendency to procure a wide variety of products, in particular rattans, timber and bamboo from secondary and mature forest as near to the village as possible, and adjacent cultivated land, leads inevitably to the depletion of more stable associations of forest the nearer one is to the main loci of settlement. Consequently, when mature forest is cut for gardens it has almost always been considerably modified already and contains plant associations more typical of regenerated secondary forest, tends to lack rattan and is considerably thinned on account of the cutting of timber for construction purposes. Such depleted but ecologically distinctive areas, such as open secondary associations often subject to marginal cultivation as well as complete clearance (Ellen 1978: 76, 85, map 9; 1999), were originally termed by Richards (1996: 379, 400) *depleted* forest and by Fosberg *altered* forest (1962: 257). Ecologically, these contain a combination of the properties of both mature and secondary forest growth. It is now widely acknowledged that the edges of garden land, swidden regrowth and disturbed and other secondary forest commonly represent the most important patches for hunting (e.g. Linares 1976), and sites of intensive extraction of plant extraction (Grenand 1992).

Nuaulu Terms and Categories Applied to Forest: Concepts and Plots

Nuaulu categorisation and general understanding of forest reflect: (a) disturbance history, (b) topography and substrate, and (c) salient species associations, nuanced in terms of (d) land ownership and (e) toponyms. I argue elsewhere (Ellen, unpubl.) that this is a broad and flexible framework, which although employing a limited set of fixed and shared lexically-labelled concepts, accommodates knowledge of wide-ranging ecological differences in forest type, and indeed constitutes a pragmatic response to the recognition of its complexity. The Nuaulu term *wesie* broadly indicates all forest, but narrowly and prototypically is understood as mature forest, far away from human settlement, which has not been modified in recent times. Once cut, individual areas of cultivation are known as *nisi*, 'gardens', which in turn can be divided into three basic types: (1) *nisi honue*, recently cleared garden plots up until the end of the first year; (2) *nisi monai*, gardens after the first year; and (3) *nisi ahue*, secondary growth of various kinds. One special category based on disturbance history is indicated by the term *nia monai* (literally, 'old village'), which refers to an old village that is still inhabited, but also to old village sites at different stages following abandonment.

In addition to disturbance history, Nuaulu describe forest locations in terms of four categories based on topography and altitude (Ellen 1978: 114, map 10): (1) *watane*: flat areas (the coastal margins, valley floors, alluvium); (2) *sanene*: valley sides; (3) *pupue*: ridge land, crests, the higher reaches of valley walls: (4) *tinete, pupue tinete*: mountains, peaks; or combinations of these terms, sometimes in conjunction with some reference to their underlying substrates. Shared labelled categories referring to areas dominated by a particular species are rare in Nuaulu, though any patch where a particularly salient species is dominant may be described with a term such as *wesi mukune* (tree-fern forest), *wesi iane* (*Canarium commune* forest), or *oni-oni* (*Cylindrica exaltata*, alang-alang grassland), but we should not mistake these for fixed terms, even though their use evokes widely shared meanings and knowledge. Where there are special terms these tend to be for deliberate anthropogenic patches, groves or plantations. Thus, strictly ecological criteria elide with cultivation (*nisi*) and ownership (*wasi*) categories. A special case of tenure which intrudes into the lexicon to describe different kinds of forest is *sin wesie*, areas of sacred protected forest (e.g. plot 7). These are not necessarily historically undisturbed or unmanaged, but are generally ecologically mature, resource rich and with a composition which reflects their age and successional stage in the development of long fallow.

These cross-cutting ethnoecological and social categories are integrated and articulated through a detailed toponymic grid. No description of forest can make much sense for Nuaulu without such an annotation. The main components of this grid are named rivers, even small creeks, supplemented by names of peaks, hills, prominent rock outcrops, stones, waterfalls, lakes, swamps, caves and suchlike. In addition there are the transient features – large trees, paths, log bridges, burned patches, patches of grassland; plus the recorded evidence of human activity, gardens belonging to particular individuals, old gardens, abandoned gardens and – most importantly – old village sites, and sites of some other special significance, such as Kamnanai Ukune or Nusi Ukune, in these cases within the sago forest at Somau. The extensive character of participatory mapping exercises elsewhere has shown just how detailed this knowledge is. On the whole, as a reference system, these toponyms begin with the names of particular mountains on the one hand, and large rivers on the other. The mountains or hills are fixed points which also give their names to large areas surrounding them. Similarly, the large rivers indicate extensive riparian and valley areas rather than the rivers themselves. Linking a river name to a mountain, therefore, provides some general coordinates, which can then be refined further by referring to tributaries of the main rivers, and tributaries of tributaries. Only when this set of coordinates are insufficient to locate places will other indicators be introduced. This set of toponyms serves to identify particular patches of forest, which to some extent bypasses the need to identify forest in terms of floristic composition or habitat structure. The toponymic references clearly indicate the investment of history in the description of a particular landscape, no better revealed than through the narrative associations of old village sites. No stretch of vegetation is ever seen as an example of some generic ahistoric type, but rather as a place whose character must be understood through its particular historical associations, and the overall 'cultural density' (Brosius 1986: 175) of the landscape.

If we now look at the plot descriptions in terms of the words Nuaulu use to describe their overall character (Table 1), five are described as *wesie* (forest), two as *nisi ahue* (long-term fallow), two as *niamonai* (old village sites) and one as *sin wesie* (sacred forest). Only a small number of terms are consistently shared by Nuaulu to describe forest habitats, and there is a low degree of lexicalisation compared with what we find in official and scientific classifications, and indeed in the folk classifications of some Amazonian peoples (Ellen, unpubl.). Systematic data on ecological knowledge and linguistic evidence indicate: (1) that the categories *wesie* (forest) and *nisi ahue* (long fallow) absorb a great deal of variation; (2) that disturbance history is the main and unifying basis for local understandings of variation, modified by occasional considerations of topography and substrate; (3) that forest is perceived as being in a constant state of flux, in large part due to interaction with humans; (4) that some stable categories are associated with specific species, but that named categories of this kind are rare; and (4) that ethnoecological understandings of forest are inseparable from categories dividing forest in terms of patterns of ownership and the cultural division of landscape reflected in the use of toponyms.

Discussion

Scientific and folk classifications have coevolved in recent global history, and the relationship between folk knowledge and instituted scientific knowledge can be modelled as two interacting and mutually reinforcing streams: hybridising through mutual borrowings while maintaining permeable boundaries for social and professional reasons, and in the interests of cognitive efficiency (Ellen and Harris 2000; Ellen 2004; Dove et al., this volume). Because tropical forest ecology and forestry are field-based practices, they have absorbed more from local knowledge systems than the other way around, and also compared with some other sciences. Indeed, instituted professional forestry has adopted much from local artisanal forestry practices, nomenclatures and understandings. The precise form this has taken varies from one country to the next, but colonial forestry services certainly appropriated much from indigenous knowledge. Thus, the Brunei Forestry Service today utilises a typology developed by colonial foresters operating in Malaya, Sarawak and British North Borneo (Kathirithamby-Wells 2004) in which 'Peat Swamp Forest' is sub divided using vernacular Malay terminology into, amongst other categories, *atan bunga* and *padang atan* (forms of forest dominated by *Shorea albida*), *padang keruntum* (dominated by *Combretocarpus rotundatus*) and '*padang* forest'; in addition to utilising the category *kerangas* (tropical heath forest) (Brunei 1984). Whatever these terms once meant, they now reflect official categories.

A similar process of knowledge transfer can be identified in colonial map-making traditions. Thus, the maps produced by the Dutch Topographische Dienst in 1917 of Seram, which surveyed in detail the entire island for the first time, show evidence of extensive reliance on Nuaulu (and other indigenous) topographic knowledge, through the use of recognisably Nuaulu toponyms over a large swathe of the central part of the island, approximately corresponding to that area which Nuaulu clans claim as their territory today. These descriptions were obviously generated by the map makers surveying with Nuaulu guides in the first decade of the twentieth century. Conducting research on these issues from 1970 onwards, and particularly in 1996, both Nuaulu co-researchers and myself have been struck by the congruence between current Nuaulu knowledge, as indicated in culturally annotated sketch-maps which Nuaulu produced for me (c.f.Fernandez-Gimenez 1993), and the toponyms provided in the 1917 Dutch map. I have already indicated in the preceeding section how crucial local toponymic knowledge is in providing a framework for understanding vegetational diversity more generally, and it is certainly not a coincidence that as *field*-based practices, colonial cartography and forestry converge in the way they made use of local knowledge.

But while colonial forestry and cartography depended heavily on the inputs of local people, at the same time there was increasing pressure to produce generic typologies of practical value to science and industry which applied over wider geographic areas. The possibilities permitted by new technologies of literary and graphic representation, in addition to the requirement to confirm qualitative intuitions with quantitative measurement, accelerated this tendency: routinising the use of plot surveys and yielding increasingly complex and contrastive typologies of forest habitats, but also raising issues of comparability between timber-type maps using different categories in different places (Avery and Burkhart 1994: 262). Sometimes the process of generalisation encouraged dangerous distortations of local ecological realities, which served to reinforce the 'territorialization strategies' and official prejudice about local forestry practices and their consequences (e.g. Fairhead and Leach 1996). More recent technologies of GIS and remote sensing have had a similar effect, creating new opportunities for distancing official and local representations. The history of attempts by ecologists and foresters to impose overly rigid classifications is reminiscent of Campbell's (2002) instructive demonstration of how the predetermined conceptual assumptions and technical specifications of a GIS package prevented the absorption of relevant data on Namibian agropastoralist land tenure, which, just like Nuaulu forest knowledge and tenure, is informal, flexible and overlapping.

Recent scientific modelling of rainforest in terms of a complex mosaic rather than as an aggregation of discrete types has been a response to the problems generated by the mechanical application of these methodologies and the assumptions associated with stereotypical, overgeneralised and essentialised representations. The diversity patterns we can now read into tropical rainforest make classifications of forest 'types' difficult. All lowland tropical rainforest is heavily influenced by patterns of human settlement and extraction over many thousands of years, and essentialist descriptions which ignore human disturbance are now widely acknowleded as misleading. And paradoxically, the problems of 'groundtruthing' imagery based on remote sensing have led to the revision of just how such data should be interpreted, in some cases involving participatory mapping. We now appreciate much more how grouping secondary biodiversity at different levels of geographic aggregation may result in very different classificatory patterns and require different kinds of analysis and interpretation (Moran 1990), and how much professional forestry can learn from local people (Wiersum 2000). Indeed, what success recent strategies, such as Joint Forest Management (JFM), have had has been grounded in the partial resolution of the opposition between global scientific forestry knowledge and local knowledge (Sivaramakrishnan 2000: 61). Even in Indonesia, the rethinking of forest policies, in the light of the failure of top-down models, made possible by the 'Reformasi' following the end of the Suharto regime in 1998, has given more scope for social forestry, local voices, and for the recognition of local community rights and ecological knowledge.

We can observe, therefore, a convergence between how local tropical forest dwellers perceive and classify forest, in this case the Nuaulu, and how scientists have reacted to the inadequacies of an earlier generation of models. The spatial variation of secondary biodiversity (habitats, biotopes, ecosystems) must, on the whole, be understood very differently from variation at the species level. My data support the claim that a classificatory model composed of large numbers of forest types, analogous to folk taxonomic schemes reported for individual plant species and typified by morphological discontinuity, does not reflect accurately how Nuaulu perceive and encode forest differences. Rather, Nuaulu representation of forest is non-taxonomic, constructed on the basis of the intersection of a number of classificatory dimensions based on different criteria acknowledging its continuous variation, deployed in a flexible and non-mechanical way. Terminologies arising from classificatory stimuli are more likely to be ad hoc descriptions of difference rather than indicating the presence of widely shared and fixed categories. Nuaulu, therefore, seem to experience forest in the same way as those ecologists who have tried to use plots to measure compositional and structural diversity. To describe a 'patch' in terms of a permanent and simple ethnoecological category is to overgeneralise and reify in a way which is not always helpful to the representation and management of resources.

Acknowledgements

All Nuaulu fieldwork reported here (1970–2003) has been sponsored by the Indonesian Institute of Sciences (LIPI: Lembaga llmu Pengetahuan Indonesia) and latterly also by Pattimura University, Ambon. The 1996 phase was funded from ESRC grant R000 236082, 'Deforestation and Forest Knowledge in South Central Seram, Eastern Indonesia', and the completion of the chapter was possible through ESRC grant RES-000-22-1106. For technical assistance, I would like to thank Dr Johannis Mogea of the Herbarium Bogoriense, Dr Johan Iskandar of the Institute of Ecology, Padjadjaran University, Bandung, and Dr Helen Newing. Christine Eagle and Lesley Farr helped with the maps, plot diagrams, and with the development of the ethnobotanical database, while Simon Platten and Amy Warren have assisted with data processing. A version of this paper was delivered at the Eighth International Congress of Ethnobiology held in Addis Ababa in 2002, and I am indebted to the British Academy for enabling me to attend this meeting. As ever, I rely on the continuing and enthusiastic involvement and support of Nuaulu friends and co-researchers in Rohua.

Note

1. A patchy distribution pattern is one in which values, observations or individuals are more aggregated or clustered than in a random distribution, indicating that the presence of one individual or value increases the probability of another occurring nearby (Lincoln et al. 1982). Alternatively, it might be defined as heterogeneity in the distribution of resources and of the patches themselves. In the context of forest ecology, patchiness reduces the possibility of accurate mapping using a neat classification of forest types.

References

Allan, C.L. 2002. 'Amerindian Ethnoecology, Resource Use and Forest Management in Southwest Guyana'. Unpublished Ph.D. thesis. University of Surrey (Roehampton).

Anon 1917. 'Boschwezen'. *Encyclopaedie an Nederlansch-Indië*, 2nd edition, Part 1, A–G., pp. 385–92. Leiden: Brill, Gravenhage: Nijhoff.

Anderson, A.B. and D.A. Posey 1989. 'Management of a Tropical Scrub Savanna by the Gorotire Kayapó of Brazil', in *Resource Management in Amazonia:*

Indigenous and Folk Strategies, ed, D.A. Posey and W. Balée, Advances in Economic Botany 7: 159–73.

Audley-Charles, M.G. 1993. 'Geological Evidence Bearing upon the Pliocene Emergence of Seram, an Island Colonizable by Land Plants and Animals', in *Natural History of Seram, Maluku, Indonesia,* ed, I.D. Edwards, A.A. MacDonald and J. Proctor, pp. 13–18. Andover: Intercept.

Aumeeruddy, Y. and J. Bakels 1994. 'Management of a Sacred Forest in the Kerinci Valley, Central Sumatra: an Example of Conservation of Biological Diversity and Its Cultural Basis', *Journal d'Agriculture Tropicale et de Botanique Appliquée* 36(2): 39–65.

Avery, T.E. and H.E. Burkhart 1994. *Forest Measurements.* New York: McGraw-Hill.

Balée, W. 1989. 'The Culture of Amazonian Forests', in *Resource Management in Amazonia,* (ed.) D.A. Posey and W. Balée, Advances in Economic Botany 7: 1–21.

———— 1992. 'People of the Fallow: a Historical Ecology of Foraging in Lowland South America', in *Conservation of Neotropical Forests: Working from Traditional Resource Use,* ed, K.H. Redford, and C. Padoch, pp. 35–57. New York: Columbia University Press.

———— 1994. *Footprints of the Forest: Ka'apor Ethnobotany: the Historical Ecology of Plant Utilization by an Amazonian People.* New York: Columbia University Press.

Bernstein, J., R.F. Ellen and Bantong Antaran 1997. 'The Use of Plot Surveys for the Study of Ethnobotanical Knowledge: a Brunei Dusun Example', *Journal of Ethnobiology* 17(1): 69–96.

Blumler, M.A. 1996. 'Ecology, Evolutionary Theory and Agricultural Origins', in *The Origins and Spread of Agriculture in Eurasia,* (ed.) D.R. Harris, pp. 25–50. London: UCL Press.

Boom, B.M. 1989. 'Use of Plant Resources by the Chácobo', in *Resource Management in Amazonia: Indigenous and Folk Strategies,* (eds), D.A. Posey and W. Balée, Advances in Economic Botany 7: 78–96.

Boomgard, P. 1994. 'Colonial Forest Policy in Java in Transition, 1865–1916', in *The Late Colonial State in Indonesia: Political and Economic Foundations of the Netherlands Indies, 1880–1942,* (ed.) R. Cribb, pp. 117–37. (Verhandelingen an het Koninklijk Instituut oor Taal-, Land- en Volkenkunde 163) Leiden: KITLV Press.

Brookfield, H.C., L. Potter and Y. Byron 1996. *In Place of the Forest: Environmental and Socio-economic Transformation in Borneo and the Eastern Malay Peninsula.* Tokyo: United Nations University Press.

Brosius, J.P. 1986. 'River, Forest and Mountain: the Penan Gang Landscape', *Sarawak Museum Journal* 36(57) (New Series): 173–84.

Brunei 1984. *Brunei Forest Resources and Strategic Planning Study, Forest-type Map. Sheet 4.* Singapore: Anderson and Marsden.

Campbell, J.R. 2002. 'Interdisciplinary Research and GIS: Why Local and Indigenous Knowledge are Discounted', in *Participating in Development: Approaches to Indigenous Knowledge*, (eds), P. Sillitoe, A. Bicker and J. Pottier, pp. 189–205. London and New York: Routledge (ASA Monograph Series No. 39).

Clay, J.W. 1988. *Indigenous Peoples and Tropical Forests: Models of Land Use and Management.* New York: Cultural Survival Inc.

Colfer, C.J.P., N. Peluso and S.C. Chin 1997. *Beyond Slash and Burn: Building on Indigenous Management of Borneo's Tropical Rainforests.* New York: The New York Botanical Garden.

Condit, R. 1996. 'Defining and Mapping Vegetation Types in Mega-diverse Tropical Forests', *Trends in Ecology and Evolution* 11: 4–5.

Departemen Kehutanan 1986. *Sejarah Kehutanan Indonesia.* 2 vols. Jakarta: Menteri Kehutanan.

Dove, M.R. 1983. 'Theories of Swidden Agriculture and the Political Economy of Ignorance', *Agroforestry Systems* 1(3): 85–99.

Edwards, I.D. 1994. 'Rainforest People', *Between the Leaves: the DPI Forest Service Journal.* Brisbane: Queensland Forest Service, Summer.

Edwards, I.D., R.W. Payton, J. Proctor and S. Riswan 1990. 'Altitudinal Zonation of the Rain Forests in the Manusela National Park, Seram, Maluku, Indonesia', in *The Plant Diversity of Malesia*, (eds), P. Baas, K. Kalkman and R. Geesink, pp. 161–75. Amsterdam: Kluwer.

———— 1993. 'Rainforest Types in the Manusela National Park', in *Natural History of Seram, Maluku, Indonesia*, (eds), I.D. Edwards, A.A. MacDonald and J. Proctor, pp. 63–74. Andover: Intercept.

Ellen, R.F. 1978. *Nuaulu Settlement and Ecology: the Environmental Relations of an Eastern Indonesian Community.* The Hague: Martinus Nijhoff. (Verhandelingen van het Koninklijk Instituut voor Taal-, Land- en Volkenkunde No. 83).

———— 1985 'Patterns of Indigenous Timber Extraction from Moluccan Rain Forest Fringes', *Journal of Biogeography* 12: 559–87.

———— 1988. 'Foraging, Starch Extraction and the Sedentary Lifestyle in the Lowland Rainforest of Central Seram', in *Hunters and Gatherers: History, Evolution and Social Change*, (eds), T. Ingold, D. Riches and J. Woodburn, pp. 117–34. London: Berg.

———— 1999. 'Forest Knowledge, Forest Transformation: Political Contingency, Historical Ecology and the Renegotiation of Nature in Central Seram', in *Transforming the Indonesian Uplands: Marginality, Power and Production*, (ed.) Tania Li, pp. 131–57. Amsterdam: Harwood.

———— 2003. 'A Synoptic View of the Co-management of Natural Resources', in *Co-management of Natural Resources in Asia: a Comparative Perspective*, (eds), G.A. Persoon, D.M.E. van Est and P. Sajise, pp. 281-92. NIAS Press: Copenhagen, Denmark.

———— 2004. 'From Ethno-science to Science, or "What the Indigenous Knowledge Debate Tells Us about How Scientists Define Their Project"', *Journal of Cognition and Culture* 4: 409–50.

———— (2001). 'Nuaulu Knowledge of Sago Palm (*Metroxylon sagu* Rottboell) Diversity in South Central Seram, Maluku, Eastern Indonesia'. Lecture delivered at Yale University Council on Southeast Asia Area Studies, 28 November.

———— (unpubl.). 'Why Aren't the Nuaulu Like the Matsigenka? Knowledge and Categorisation of Forest Diversity on Seram, Eastern Indonesia'. Paper delivered at the Eighth Internatioinal Congress of Ethnobiology held in Addis Ababa, September 2002.

Ellen, R.F. and H. Harris. 2000. 'Introduction', in *Indigenous Environmental Knowledge and its Transformations: Critical Anthropological Perspectives*, (eds), R. Ellen, P. Parkes and A. Bicker pp. 1–33. Amsterdam: Harwood.

Eyre, F.H. (ed.) 1980. *Forest Cover Types of the United States and Canada*. Washington DC: Society of American Foresters.

Fairhead, J. and M. Leach 1996. *Misreading the African Landscape: Society and Ecology in a Forest-savanna Mosaic*. Cambridge: Cambridge University Press.

———— 1998. *Reframing Deforestation: Global Analysis and Local Realities – Studies in West Africa*. London: Routledge.

Fernandez-Gimenez, M. 1993. 'The Role of Ecological Perception in Indigenous Resource Management: a Case Study from the Mongolian Forest-steppe', *Nomadic Peoples* 33: 31–46.

Fleck, D.W. and J.D. Harder 2000. 'Matses Indian Rainforest Habitat Classification and Mammalian Diversity in Amazonian Peru', *Journal of Ethnobiology* 20(1): 1–36.

Fosberg, F.R. 1962. 'Nature and Detection of Plant Communities Resulting from Activities of Early Man', in *Symposium on the Impact of Man on Humid Tropics Vegetation, Goroka, Territory of Papua and New Guinea*, 1960, pp. 251–52. Djakarta: UNESCO Science Cooperation Office for Southeast Asia.

Glatzel, G. 1992. *Report of a Fact Finding Mission on the Proposal for a Centre for Sustainable Land Use on Forested Islands of the Moluccas*. Vienna: Bundeskanzleramtes der Republik Österreich.

Greig-Smith, P. 1964. *Quantitative Plant Ecology*. London: Butterworth.

Grenand, P. 1992. 'The Use and Cultural Significance of the Secondary Forest among the Wayapi Indians', in *Sustainable Harvest and Marketing of Rain Forest Products*, (eds), M. Plotkin and L. Famolare, pp. 27–40. Washington DC: Island Press.

Howard, J.A. 1991. *Remote Sensing of Forest Resources: Theory and Application.* London: Chapman and Hall.

Johns, R.J. 1990. 'The Illusory Concept of the Climax', in *The Plant Diversity of Malesia*, (eds), P. Baas, K. Kalkman and R. Geesink, pp. 133–46. The Netherlands: Kluwer.

Johnston, C.A. 1998. *Geographic Information Systems in Ecology.* Oxford: Blackwell Science.

Johnston, M. 1998. 'Tree Population Studies in Low-diversity Forests, Guyana. II. Assessments on the Distribution and Abundance of Non-timber Forest Products', *Biodiversity and Conservation* 7: 73–86.

Kathirithamby-Wells, J. 2004. *Nature and Nation: Forests and Development in Peninsular Malaysia.* Honolulu: University of Hawaii.

Kent, M. and P. Coker 1992. *Vegetation Description and Analysis.* London: Bellhaven Press.

Kershaw, K.A. 1973. *Quantitative and Dynamic Plant Ecology.* London: Arnold.

Latinis, D.K. 2000. 'The Development of Subsistence System Models for Island Southeast Asia and Near Oceania: the Nature and Role of Arboriculture and Arboreal-based Economies', *World Archaeology* 32(1): 41–67.

Linares, O.F. 1976. '"Garden hunting" in the American Tropics', *Human Ecology* 4(4): 331–49.

Lincoln, R.J., G.A. Boxshall and P.F. Clark 1982. *A Dictionary of Ecology, Evolution and Systematics.* Cambridge: Cambridge University Press.

Maloney, B.K. 1993. 'Climate, Man, and Thirty Thousand Years of Vegetation Change in North Sumatra', *Indonesian Environmental History Newsletter* 2: 3–4.

Martin, G.J. 1995. *Ethnobotany: a Methods Manual.* London: Chapman and Hall.

Moran, E.F. 1990. 'Levels of Analysis and Analytical Level Shifting: Examples from Amazonian Ecosystem Research', in *The Ecosystem Approach in Anthropology: from Concept to Practice* (2nd edition), (ed.) E.F. Moran, pp. 279–308. Ann Arbor: The University of Michigan Press.

Muraille, B. 2000. 'Associating People and Industrial Forestry: Joint Forest Management in Lao PDR', in *Forestry, Forest Users and Research: New Ways of Learning*, (ed.) A. Lawrence, pp. 71–84. Wageningen, The Netherlands: European Tropical Forest Research Network.

Padoch, C. and N. Peluso. 1996. 'Changing Resource Rights in Managed Forests of West Kalimantan', in *Borneo in Transition: People, Forests, Conservation and Development*, (eds), C. Padoch and N. Peluso, pp. 121–36. Kuala Lumpur: Oxford University Press.

Padoch, C. and C. Peters 1993. 'Managed Forest Gardens in West Kalimantan, Indonesia', in *Perspectives on Biodiversity: Case Studies of Genetic Resource Conservation and Development*, (eds), C. Potter, J. Cohen and D. Janczewski, pp.167–76. Washington D.C.: AAAS.

Paijmans, K. 1970. 'An Analysis of Four Tropical Rainforest Sites in New Guinea', *Journal of Ecology* 58: 77–101.

Peluso, N.L. 1992. *Rich Forests, Poor People: Resource Control and Resistance in Java*. Berkeley: University of California Press.

——— 1996. 'Fruit Trees and Family Trees in an Anthropogenic Forest: Ethics of Access, Property Zones, and Environmental Change in Indonesia', *Comparative Studies in Society and History* 38: 510–48.

Persoon, G., T. Minter, B. Slee and C. van der Hammer. 2004. *The Position of Indigenous Peoples in the Management of Tropical Forests*. Wageningen, the Netherlands: Tropenbos International (Tropenbos Series 23).

Pires, J.M. and G.T. Prance. 1985. 'The Vegetation Types of the Brazilian Amazon', in *Key Environments: Amazonia*, (eds), G.T. Prance and T.E. Lovejoy, pp. 109–45. Oxford and New York: Pergamon Press – IUCN.

Poulsen, A.D., I.C. Nielsen, S. Tan and H. Balslev. 1996. 'A Quantitative Inventory of Trees in one Hectare of Mixed Dipterocarp Forest in Temburong, Brunei Darussalam', in *Tropical Rainforest Research – Current Issues*, (eds), D.S. Edwards, W.E. Boloth and S.C. Choy, pp. 139–50. Dordrecht: Kluwer.

Proctor, J., J.M. Anderson, P. Chai and H.W. Vallack. 1983. 'Ecological Studies in Four Contrasting Rainforests in Gunung Mulu National Park, Sarawak. 1. Forest Environment, Structure and Floristics', *Journal of Ecology* 71: 237–60.

Puri, R.K. 2005. 'Post-abandonment Ecology of Penan Forest Camps: Anthropological and Ethnobiological Approaches to the History of a Rainforested Valley in East Kalimantan', in *Conserving Nature in Culture: Case Studies from Southeast Asia*, (eds), M.R. Dove, P. Sajise, and A. Doolittle. New Haven: Yale University Council on Southeast Asia Studies.

Rambo, A.T. 1979. 'Primitive Man's Impact on Genetic Resources of the Malaysian Tropical Rainforest', *Malaysian Applied Biology* 8(1): 59–65.

Richards, P.W. 1996 (1952). *The Tropical Rain Forest*. Cambridge: Cambridge University Press.

Salick, J. 1989. 'Ecological Basis of Amuesha Agriculture, Peruvian Upper Amazon', in *Resource Management in Amazonia: Indigenous and Folk Strategies*, (eds), D.A. Posey and W. Balée. Advances in Economic Botany 7: 189–212.

Schmink, M., K.H. Redford and C. Padoch 1992. 'Traditional Peoples and the Biosphere: Framing the Issues and Defining the Terms', in *Conservation of Neotropical Forests: Working from Traditional Resource Use*, (eds), K.H. Redford and C. Padoch, pp. 3–13. New York: Columbia University Press.

Scott, J.C. 1998. *Seeing Like a State: How Certain Schemes to Improve the Human Condition have Failed*. New Haven, CA and London: Yale University Press.

Shepard, G.H., D.W. Yu, M. Lizarralde and M. Italiano. 2001. 'Rainforest Habitat Classification among the Matsigenka of the Peruvian Amazon', *Journal of Ethnobiology* 21(1): 1–38.

Shepard, G.H., D.W. Yu and B.W. Nelson 2004. 'Ethnobotanical Ground-truthing and Forest Diversity in the Western Amazon', in *Ethnobotany and Conservation of Biocultural Diversity*, (eds), L. Maffi, T. Carlson and E. Lopez-Zent, pp. 133–71. New York: New York Botanical Gardens. (Advances in Economic Botany, vol. 15.)

Sidiyasa, K. and I.G.M. Tantra 1984. 'Tree Flora Analysis of the Way Mual Lowland Forest, Manusela National Park, Seram-Maluku', *Bulletin Penelitian Hutan* 462: 19–34.

Sillitoe, P. 1996. *A Place against Time: Land and Environment in the Papua New Guinea Highlands.* Amsterdam: Harwood. (Studies in Environmental Anthropology 1.)

———— 1998. 'An Ethnobotanical Account of the Vegetation Communities of the Wola Region, Southern Highlands Province, Papua New Guinea', *Journal of Ethnobiology* 18(1): 103–28.

———— 2002a. 'Participant Observation to Participatory Development: Making Anthropology Work', in *Participating in Development: Approaches to Indigenous Knowledge*, (eds), P. Sillitoe, A. Bicker and J. Pottier, pp. 1–23. London and New York: Routledge. (ASA Monograph Series No. 39.)

———— 2002b. 'Globalizing Indigenous Knowledge', in *Participating in Development: Approaches to Indigenous Knowledge*, (eds), P. Sillitoe, A. Bicker and J. Pottier, pp. 108–38. London and New York: Routledge. (ASA Monograph Series No. 39.)

Sivaramakrishnan, K. 2000. 'State Sciences and Development Histories: Encoding Local Forestry Knowledge in Bengal', *Development and Change* 31(1): 61–90.

Spencer, J.E. 1966. *Shifting Cultivation in Southeastern Asia.* Berkeley and Los Angeles: University of California Press.

Sprugel, D.G. 1991. 'Disturbance, Equilibrium and Environmental Variability: What is "Natural" Vegetation in a Changing Environment?', *Biological Conservation* 58: 1–18.

Suharno, D.M. 1997. 'Representation de l'environment vegetal et pratiques agri-coles chez les Alune de Lumoli, Seram de l'ouest (Moluques Centrales, Indonesie de l'Est)', These de Doctorat de l'Université Paris VI.

Torquebiau, E.F. 1987. 'Forest Mosaic Pattern Analysis', in *Report of the 1982–1983 Bukit Raya Expedition*, (ed.) H.P. Nooteboom pp. 25–42. Leiden: The Rijksherbarium.

Valencia, R., H. Balslev and G. Paz Y Miño 1994. 'High Tree Alpha-diversity in Amazonian Ecuador', *Biodiversity and Conservation* 3: 21–28.

Vandergeest, P. and N.L. Peluso 1995. 'Territorialization and State Power in Thailand', *Theory and Society* 24: 385–426.

Whitmore, T.C. 1984. *Tropical Rain Forests of the Far-east.* Oxford: Clarendon Press.

Whitmore, T.C., K. Sidayasa and T.J. Whitmore 1987. 'Tree Species Enumeration of 0.5 Hectare on Halmahera', *Garden Bulletin Singapore* 40: 31–34.

Whitmore, T.C., I.G.M. Tantra. and U. Sutisna. (eds) 1989. *Tree Flora of Indonesia: Checklist for Maluku.* Bogor: Ministry of Forestry, Agency for Forestry Research and Development.

Wiersum, K.F. 2000. 'Incorporating Indigenous Knowledge in Formal Forest Management: Adaptation or Paradigm Change in Tropical Forestry', in *Forestry, Forest Users and Research: New Ways of Learning,* (ed.) A. Lawrence, pp. 19–32. Wageningen, The Netherlands: European Tropical Forest Research Network.

4 'Indigenous' and 'Scientific' Knowledge in Central Cape York Peninsula

Benjamin R. Smith

During my doctoral fieldwork in central Cape York Peninsula I spent a considerable amount of time moving between the township of Coen and a number of 'outstations' in its hinterland. ('Outstations' are small camps established by family groups on land with which they have 'traditional' ties). The men and women with whom I was working also used their frequent journeys between Coen and outstations to visit other important places en route. One such place – visited only by men – was referred to as a '*chemist*['s] *shop*' in local Murri (Aboriginal) English.[1] Traditional plant '*medicines*' were gathered at this place and brought back into town. These '*medicines*' were understood to effect '*luck*' in a series of town-based activities, including gambling at card games and the pursuit of sexual liaisons.[2]

The use of the term '*chemist*['s] *shop*' in relation to such local knowledge suggests a willingness amongst the Peninsula's indigenous population to draw parallels between the artefacts of 'indigenous knowledge' and products of 'scientific knowledge'. In this way, the metaphor of the chemist's shop seems to support the use of the synonym 'local science' for local knowledge. But other events during my fieldwork make it clear that the relationship between local knowledge and science is more ambivalent.

At a public meeting in Coen, for instance, scientists employed by the Queensland Parks and Wildlife Service (QPWS) sought to inform local people about a media scare over the Equine Morbillivirus (now called the Hendra virus), which was thought to be able to cross from flying foxes to humans. Red flying foxes (*minh wuki* in the Mungkanhu language) continue to be harvested from colonies around Coen by local Aborigines, who knock them from trees while they roost. QPWS intended to warn people of the danger of being scratched or bitten by an infected animal, whilst downplaying possible fears about the disease generated by a media scare (assuring people, for instance, that the meat remained safe to eat). This meeting followed the usual pattern of such events, in which visiting '*whitefellas*' were met with blank looks by the Aboriginal audience, many of whom were alienated by their use of unfamiliar language – '*rich English*' and '*big words*'. But I sensed a particular reaction – a mixture of incredulity and amusement – when one scientist explained that the large colony of fruit bats currently roosting in Coen had flown up to Coen from Ravenshoe, a town several hundred kilometres to the south. My suspicions were confirmed when one of the older men present at the meeting (also one of the family on whose 'traditional land' the 'chemist shop' was situated) came up to me and wryly asked me whether I thought he 'should've told 'em where flying foxes *really* come from'.[3]

For many – if not all – of the Aborigines living in central Cape York Peninsula, flying foxes continue to be associated with a body of local knowledge pertaining to '*rainbow serpents*'. These powerful creator beings are held to remain active across the region, and are particularly associated with watercourses and deep, permanent pools of fresh water ('*lagoons*'). These '*rainbows*' are said to be roused by the '*smell*' of people with no ties to a given area or '*country*', or by actions that contravene local customary rules or '*Murri law*' (see Merlan 2000, 2005). Flying foxes are closely associated with rainbows, which are said to swallow them, carry them in their mouths and then release them elsewhere, the flying foxes bubbling up and bursting from the surface of the water. Locally, this is where 'flying foxes *really* come from', and the reserve of the local Aboriginal attendees at the QPWS meeting, and the amusement of the man who approached me afterwards indicate a reticence to broach a perceived gap between 'scientific knowledge' and 'local knowledge' of this kind.[4] Indeed, the same wry humour that characterised this man's (rhetorical) question to me – and, doubtless, his sense that I was stranded somewhere between the QPWS scientists' 'rational' explanation and local Murri knowledge of flying foxes – underlay the use of a '*white name*' linked with science ('*chemist*['s] *shop*') for the practical use of another aspect of local knowledge.

The aim of this paper is to respond to the ways in which anthropologists and others have encouraged the anthropological recognition of 'local knowledge' as 'local science' (a suggestion that has also built upon the sociological critique of the idea and practice of science). As with anthropological accounts of local knowledge more generally, such claims are most often made in the context of knowledge pertaining to local environments and their use and management. These claims also commonly relate to 'development', its subjects and its 'victims'. But whilst the concept of local science provides a powerful metaphor, ensuring greater attention to and respect for local knowledge, I want to suggest that glossing local knowledge as 'local science' may (in some instances) obscure important particularities of local knowledge systems and the broader social and cultural contexts within which they have developed.

The basis for my argument is akin to Lévi-Strauss's observation that 'thought, and the world which encompasses it, are two correlative manifestations of the same reality' (Lévi-Strauss 1981: 678).[5] Examining the interaction of local knowledge and 'global' science in a contemporary indigenous life-world, I argue that the term 'local science' may be more problematic than it initially seems precisely because of this correlative relationship between thought – or knowledge – and world. Nonetheless, disjunctures between local knowledge and science do not necessarily hinder mutually beneficial interactions between Aborigines and scientists, or the bodies of knowledge on which they draw. Indeed, relationships of this kind have formed the basis for a number of successful 'development' projects at 'outstations' and on the indigenous homelands that surround them; these projects have been based on 'conversations' between local knowledge and scientific research. Such projects nonetheless raise questions about the ways in which '(global) science' constructs knowledge, the forms of authority tied to knowledge, and the effects of such 'authority' on local forms of cultural production.

If I am hesitant about the term 'local science', I also want to express some reservations about the term 'indigenous knowledge'. In central Cape York Peninsula, the interactions between 'local knowledge' and 'global science' increasingly demonstrate forms of local knowledge that are profoundly 'intercultural'. (I should add that I use the term 'intercultural' here *not* in the sense of a field composed of relations between distinct entities, despite possible presumptions of social actors to the contrary, but rather to designate a field of interaction in which culturally inflected meanings and practices are (re)defined through processual interrelationships between such actors.[6]) There is little doubt that the interactions constitutive of knowledge-based development projects further intensify 'intercultural' forms of local cultural production; in doing so, these projects produce increasingly 'globalised' forms of knowledge not only about the local environment, but also about 'indigenous people' themselves.

'Local knowledge' vs. 'local science'

An extensive debate has continued amongst anthropologists and in related disciplines over the scope of the term 'science', particularly regarding its extension to include 'indigenous', 'traditional' or 'local' knowledge. Both within anthropology and in the sociology of scientific knowledge there has been a growing consensus that the limitation of the term 'science' to Western, professional science has primarily served to perpetuate a worldview that privileges Western science over other, excluded ways of knowing (see Haraway 1988; Harding 1994; Franklin 1995; Nader 1996; Harris 1998; Turnbull 2000). Much of the impetus for these critiques can be located in the influence of the work of Michel Foucault, whose work on the history of systems of thought (or 'epistemology', the analysis of the grounds of knowledge) continues to be influential. Foucault's lectures given at the Collège de France in 1975–76 open with a lengthy critique of dominant knowledges from the position of 'subjugated' knowledges (which include those knowledges now labelled 'local' or 'indigenous'). For Foucault, such knowledges include all of those disqualified as naïve, hierarchically inferior, and 'knowledges that are below the required level of erudition or scientificity' (Foucault 2004: 7).

The principal aim of much of the work undertaken within the field of cultural or sociological studies of scientific knowledge – to unsettle or critique the hierarchical relationship of what might be termed 'global' or 'Western' science with 'local knowledge' (see also Dove et al., this volume) – is one with which I have much sympathy. The immeasurably beneficial result has been that it is now 'increasingly acknowledged beyond anthropology that other people have their own effective "science" and resource use practices' (Sillitoe 1998: 223). Doubtless such a critique was not only long overdue, but also necessary in order to counteract the exploitation of local knowledges in an increasingly globalised world, for example through 'bio-prospecting' (see Brush 1993; also Clift and Bodeker, both this volume). It has also been important in helping to limit the disasters caused by the emplacement of exogenous practices, in the name of 'development', in local contexts in which a more equitable dialogue between 'local' and 'global' knowledges would almost certainly have produced far better results (see Scott 1998; Sillitoe et al. 2002; Pottier et al. 2003; Bicker et al. 2004).[7]

Whilst positive arguments for broadening the use of the term 'science' to encompass indigenous or local knowledge systems have been clearly stated, exponents of the extension of 'science' to include indigenous knowledge have paid less attention to some potential pitfalls. There are undoubtedly many 'human societies doing science or accumulating knowledge by verifying observation' and employing 'a self-conscious attitude toward knowledge and knowing that embodies curiosity with empiricism' (Nader 1996: 11, 1) beyond the institutional settings of Western science. In such instances, the notion of 'local science' – or, indeed, of science per se – might be usefully deployed. But as Nader rightly notes, drawing on Bielawski's comparison of Inuit indigenous knowledge and Arctic (global) science, regions like the Arctic reveal a radical divide between 'modern science' and 'indigenous knowledge' (Nader 1996: 21; also Sable et al., this volume). In such regions, calling local knowledge '(local) science' may prove to be a mistranslation. Further, such mistranslation may obscure important aspects of local knowledge systems that may again be compounded by the implementation of inappropriate development projects in the region. As I hope to demonstrate below, the same concerns raised by Bielawski's analysis are pertinent – although not without qualification – in central Cape York Peninsula.

A key aspect of the 'dominant' distinction between science and local knowledge is the presumption (both by scientists and by members of 'modern' publics more generally) of the particular authority of 'Western' or 'modern' science. Such presumptions are often expressed through claims that 'science' has a unique capacity for absolute or universal explanation. This notion has continued to displace local knowledge and effect considerable harm to the environments where scientific-technical interventions have occurred. Here the term 'global science' seems apposite, and it is this 'global' presumption that sociological critiques of scientific knowledge has sought to problematise. Scientists have tended to elide the gap between their 'universalist' presumptions and the set of particular, internally complex and disparate knowledges (with their roots in the European Scientific Revolution) that constitute 'science-as-it-is practised'. This elision has masked the fact that 'Western technoscience', like all other knowledge traditions, is itself a form – or assemblage of forms – of local knowledge (Turnbull 2000: 4–5). As Turnbull argues, it is this masking of the locality of the forms of scientific knowledge that has allowed scientists and others to view science as a body of 'mimetic totalising theory' (that is, as a cohesive body of theory reflecting a 'real world'), producing its 'globality'. The effect of such claims is 'simultaneously to promote and reinforce' a sense of (Western) cultural stability and to effect dominance and 'to justify the dispossession of other peoples' (Turnbull 2000: 11; also Rowse 2005).[8]

This presumption of science's general authority is tied to another problem with 'global' science, the assumption of the translatability of scientific knowledge or the appropriateness of the application of knowledge drawn from one location in another location. As Nader argues, drawing on the work of Paul Richards, 'science as universally applicable knowledge is supposed to override ecological particularism and site-specific knowledge. Science derives its power precisely because it is not confined to particularities' (Nader 1996: 12). Indeed, Nader cites White's suggestion that ecological approaches are 'on principle anti-scientific, as science at present is usually conceived and practiced', given their emphasis on the specificities of particular locales, and localised forms of knowledge relating to them (Nader 1996: 12, after White 1979: 76; also Ellen, this volume).

The problem of translatability raises questions not only for the use of the term '[local] science' as a gloss for indigenous knowledge systems, but also in relation to presumptions about the ease of 'fit' between Western or 'global' science and indigenous knowledge systems even after presumptions of hierarchical inferiority have (apparently) been dealt with. Suggestions of a potential resolution between Western science and local knowledge, based on the assumption that both sets of practice and knowledge 'relate ... to the same natural world "out there," albeit expressed in quite different idioms revealing concerns for somewhat different issues' (Sillitoe 1998: 226), might be more problematic than they might first appear – particularly where Sillitoe's inverted commas and attention to the 'symbolic associations' of local knowledge are not maintained. Suggestions of this kind seem to rest on suppositions of the universal, fundamental existence of nature, and subsequent presumptions of the availability of objective forms of knowledge that may, in fact, be situated within particular, albeit globalising, ways of understanding the world. They can obscure the fact that the 'symbolic' is a constitutive aspect of all forms of knowing (and all 'environments'), rather than a separable set of 'associations' supplementing 'real' or 'pragmatic' knowledge. What is revealed by the engagement of this kind of knowledge and practice are not simply 'concerns' in the standard meaning of the term. Rather, environmental relations involve the *constitutive* revelation of particular kinds of phenomena. In Lévi-Strauss's terms, the world is not simply 'out there', but rather, is a correlate of the form of thought brought to bear on it.

This 'correlation' of world and knowledge has been dealt with at length (and with considerably more nuance and depth of analysis than is possible here) in the work of Roy Wagner. Wagner, drawing on his fieldwork in Melanesia, provides a useful account of the differing ordering effects of Western and indigenous knowledge systems in his comparison of 'scientific' and 'Indigenous Papuan' environmental knowledge.[9] For Wagner, Western scientific approaches to human–environmental relations are founded on the separation of that which lies in the world of human artifice, and the 'natural' realm, understood as the innate component of the human–environmental totality (see also Latour 1993 [1991]). Such an approach reveals 'natural phenomena' as having 'the spontaneous and self-contained character of standing for themselves' (Wagner 1977: 394), that is, as an object world 'out there' and subject to our intervention (or distanced understanding). From this (scientific) perspective, the natural thus emerges through 'a flow of seemingly innate, differentiating transformation that is precipitated by our systematic and literal efforts at harnessing or understanding it' (Wagner 1977: 394).[10] (Here we are back to Lévi-Strauss, and the correlative interrelationship of 'thought, and the world which encompasses it').

In Wagner's 'Indigenous' knowledge system – a system that demonstrates substantial similarities to the knowledge system that seems to have been extant in central Cape York Peninsula prior to the impacts of white settlement[11] – human existence is positioned quite differently in relation to the 'environment'. The resulting perspective compels the recognition of 'a kind of immanent human essence as the object and sustaining force' of local understanding and action. This immanent essence appears both as the 'moral soul' of each person and the 'tradition' of local groups, and it provides 'the binding force of the cosmos' (Wagner 1977: 404). Here the nature/culture dualism does not hold. Rather, in the 'dwelling' perspective (Ingold 2000) that inheres within such knowledge systems, particular people and 'environments' are revealed (to those peoples who have come to know the world and themselves within the horizon of such knowledge systems) as deeply interconnected, sharing of the same substance. Here agency or subjectivity is seen as extending beyond the human; what Westerners are compelled to regard as 'objects', or as 'nature', is understood as an expression of 'immanent essence' (shaped, in central Cape York, by the originary actions of the *Stories*), and thus as both of the same kind, and fundamentally inseparable from the humans who belong to a particular locale (Smith forthcoming). In such life-worlds, any presumption of the ability to treat the natural world as 'standing reserve',[12] 'object' or 'resource', open to intervention by human agents, is deeply problematic. Rather, environmental interventions must negotiate particular and substantial interrelationships between people and other aspects of local life-worlds.

Franklin (1995: 170) similarly argues that it is the effect of 'objectification' produced within Western epistemologies that reveals 'the self-evident real' of nature. She also notes that it is this 'self-evident real' that allows for the sense of the superiority involved in the distinction between (merely) 'local' and 'global' orders of knowledge. Approaching local indigenous knowledge systems on the basis of a universal relation to a nature 'out there' is thus likely to reinforce the hegemony of 'global science' in relation to local knowledge, given the foundation of these distinct knowledge systems on particular epistemological bases (that is, within particular ways of knowing), despite attempts to engage a positively valued body of local knowledge at the level of 'content'. Such an approach also leaves open the risk of generating a sense of 'knowing better' among particular practitioners of global science (and here I would not exclude anthropologists), even within attempts to respectfully engage local knowledge (and local 'knowers') through collaborations between 'local knowledge' and 'global science'. Certainly it is this kind of hegemonic effect that leads (in part) to Aboriginal people in central Cape York holding their own knowledge of their life-worlds in reserve when engaging with those, like the QPWS scientists, who they (not unjustifiably) expect will engage with claims about rainbow serpents and flying foxes in a dismissive or patronising manner.

How then might this hegemonic character of interaction be avoided in projects which involve both 'local knowledge' and 'Western science'? One possible basis for more equitable articulations between 'science' and 'local knowledge' is outlined by Donna Haraway in her work on 'situated knowledges'. For Haraway (1988: 592), these are akin to those knowledges that have been discussed here as 'local' (including the forms of scientific knowledge that Turnbull and others have argued are just as local as indigenous knowledge systems). Haraway uses the term 'situated knowledges' to advance her argument that all forms of knowledge – in particular, those associated with 'global' scientific practice – should be cognisant of their own (necessarily local) epistemological foundations. It may be useful to add that anthropological accounts of 'indigenous' knowledge systems suggest that knowledge is already commonly self-consciously localised in this manner within these systems. In such cases, the citation and use of knowledge is commonly subject to a set of place-based rules, beliefs and practices – a strongly localised *ethics* of knowledge – that governs both environmental and interpersonal interactions. Within the indigenous knowledge systems of Cape York Peninsula, for example, one's ability to properly act (indeed, to publicly *know*) has classically been constrained within one's own relation to a particular, localised corpus of knowledge, language, subjectivity and relations to people, flora, fauna and other aspects of '*country*'. In this system, as one man put it, 'you can't just cut any tree ... got to think about [the tree's] age, right time, and right person'.

Despite her critique of its globalising thrust, Haraway nonetheless sees science as an important and beneficial (set of) way(s) of knowing. The call for situated knowledges is not a call for an absolute relativism of knowledge itself. Like many anthropologists who advocate the recognition of local knowledge, Haraway does not demand the abandonment or downplaying of scientific explanations. Rather, she suggests a more ethical deployment of knowledge, and a more careful interaction of knowledges. Scientific knowledge is not to be abandoned; rather, those practising science and relying on the knowledge it generates need to recognise what Franklin (1995: 173) calls science's 'isomorphism between representation and ontology' – its elision of (and simultaneous dependence on) a particular 'correlative' relation between thought and world.[13] This isomorphism leads to science representing both 'knowledge of the natural world expressed in naturalistic terms' and 'the procedures for obtaining that knowledge', resulting in the 'conflation of instrumental technique with the "real" it describes'. For Haraway, such self-recognition would then present the possibility of 'partial, locatable, critical knowledges [including 'science'] sustaining the possibility of webs of connections called solidarity in politics and shared conversations in epistemology' (Haraway 1988: 584). 'Shared conversations' of this kind would help to avoid the displacement of local forms of knowledge through the actions of bio-prospectors, or others who have undertaken scientific research on indigenous homelands without regard for indigenous connections to '*country*'. Perhaps more importantly, such conversations would also avoid the globalising spread and unquestioned authority of particular ways of knowing that can obscure critically important, foundational aspects of 'local knowledge' systems – including the importance of *locale* or localisation within their constitution of knowledge and its relation to social life, 'environments', and particular kinds of selves.

'Indigenous knowledge' and Science in Development: Transformation at the 'knowledge interface'?

'Shared conversations' of the kind outlined by Haraway have, in fact, already been undertaken in central Cape York Peninsula in an attempt to develop socially and environmentally sustainable economic activities. One such project is based on the distillation of oils and other plant products from a number of the traditional '*medicines*' that grow on local indigenous homelands.[14] The development of this project has involved collaborative work between local indigenous people, a regional development organisation and natural scientists.

The project, instigated by the local indigenous people on whose homelands it is based, draws on particular forms of local knowledge in conjunction with Western science. The aim of this collaboration is to develop an environmentally and socially sustainable local enterprise.[15] It demonstrates the indigenous group's success in generating what Tsing calls a 'field of attraction', within which relationships are produced between a rural community and outside experts that 'keep [the] experts coming back' (Tsing 1999, cited from Dove et al., this volume). Beyond its aim of realising a socially and environmentally sustainable source of income, the oils project is also intended to advance other local indigenous aspirations, including the cross-generational maintenance and transmission of local plant knowledge through experiential involvement with important species on traditional homelands (see Heckler, this volume; also Sable et al., this volume for another 'conversational' or 'situated' interaction between local knowledge and Western science).

The strength of the oils project is the way in which it has brokered a particular, 'situated' relationship between two fields of knowledge and practice (albeit fields that are not as 'culturally distinct' as either the indigenous or non-indigenous participants in the project presume). Within this relationship 'local knowledge' and 'Western science' have articulated productively within a particular environment to the mutual satisfaction of those involved. For Aboriginal participants this has meant that their emphasis on locality has been respected, and indeed reproduced, within an 'intercultural' field. For scientists, it has meant access to a body of local knowledge, as well as a local environment and its flora, benefiting their own scientific knowledge and practice.

There is little doubt, given the increasing interpenetration of Aboriginal and 'mainstream' Australian life-worlds, even in a supposedly 'remote' region like Cape York Peninsula, that such projects are now necessary to the reproduction of certain kinds of local knowledge (for example, the 'language names' and 'traditional uses' of local plant 'medicines'). However, it is also the case that the distinctions of 'indigenous' and 'Western' knowledge and practice have been overstated and reified within such interactions (see also Heckler, this volume; Smith 2005a). This is also a situation in which the overtly hegemonic relationship between 'global science' and 'local knowledge' has shifted somewhat. In this instance, the local imbrication of science and local knowledge has created a relationship in which a sense of authority is mutually established through the interplay of knowledges. It should be noted, however, that whilst scientific and local knowledge increasingly exist – in Cape York Peninsula, as elsewhere – in what others have called 'hybrid' relationships (see Sillitoe 1998: 226), scientific knowledge has proved far more resistant than local knowledge to epistemological transformation. The result is that local 'indigenous' fields of knowledge are now complex and multiform, whilst the knowledge produced within scientific practice has maintained epistemological continuity (and, doubtless, some sense of its own broader authority) despite changing forms of scientific practice (in particular, local collaboration).

The uptake of the term 'indigenous knowledge', and the commoditisation of local knowledge now sought by many Aboriginal people as part of their 'intellectual property rights', mark the effects of the transformative imbrication of knowledge systems. The cost of the reproduction of local knowledge via the 'knowledge interface' (Sillitoe 1998: 226) appears to be a partial transformation of the 'dwelling perspective' (in which knowledge is figured as an innate part of local 'environments') to a perspective in which knowledge becomes refigured as a commodity or object. In such circumstances, the notion of a 'knowledge interface' is misleading (at least as regards the local Aboriginal perspective) – although it is nonetheless essential in order to perpetuate a sense of interaction between 'distinct systems'. This idea of separate knowledge systems is a key 'social imaginary' in the field of development. It acts to shape particular relationships within 'indigenous' locales where originally distinct forms of knowledge (and their associated social fields) are now deeply interwoven; within these relationships local people and development professionals remain impelled to imagine their interactions as occurring between distinct 'societies' (see also Smith 2005a, 2005b).

Even the selection of the medicinal properties of plants as the focus of the oils project suggests that this project, like many others of its ilk, has arisen within a strongly intercultural field. As Dove et al. (this volume; also Bodeker, this volume) note, the idea of medicinal plants circulates as a particularly operable form of traditional environmental knowledge within the 'development industry'. The 'field of attraction' that now encompasses both 'locals' and 'experts' in Cape York Peninsula, as elsewhere, has developed on the basis of the operability of ideas of this kind, and every interaction within this field acts to deepen the intercultural nature of 'indigenous knowledge'.

Despite these transformative aspects, the deepening interculturalism associated with such projects[16] is not necessarily leading towards a (total) homogenisation of knowledge, either at the level of reified identification of knowledges, or at the level of practice and meaning. Rather, the situated interactions between local indigenous people and scientists provides one means of (partially) reproducing both local knowledge and local ways of knowing that are otherwise increasingly overwhelmed even in interactions between indigenous people. However, these items and ways of knowing are increasingly aspects of 'hybrid' or multivalent knowledges held by particular persons, and circulated through the ever-more-complex social fields in which we are all now situated.

Changes in local knowledge systems are also evident in the greater willingness of a number of Aborigines to deal with their homelands in a manner that Haraway (1988) calls 'resourcing'. This is the ability to engage with a 'natural' world or environment as a set of objects or resources available for use, for example in the extraction of timber (a relation that a number of scholars, starting with Heidegger, have argued is closely connected to modern 'global' science and technology).[17] This relation of resourcing is foreign to classical indigenous modes of reckoning relationships to homelands, which are deeply relational in character, and in which extraction is impelled to be far more limited in scope – not least through restricting who can use what and when. But recent years have seen a number of Aborigines strike deals with logging companies to extract timber from their homelands in return for cash payments (although in one recent case such a venture was successfully opposed by other Aborigines on the basis that such extraction contravened local Aboriginal 'law').[18]

Even in the case of the oils project, there are indications of a shift towards a different kind of relationship to homelands. In the description of the oils project on the website of a local indigenous organisation, the Aboriginal group involved in the project affirms that they:

> possess a broad and detailed knowledge of the ecology of our environment. Our knowledge of forest and forest products, in particular the oils and resins derived from the leaves, bark, fruit, seeds and roots of woody plants, is detailed and these resources play an important role in our health and well-being at the local level. This knowledge has been accumulated by [our] people via 'diachronic' observations of the environment and passed down from generation to generation.

The language is noteworthy: 'environment' and 'resources' for instance. If Wagner is correct in his analysis (and if I am correct in suggesting that much of Wagner's analysis is pertinent to the 'classical' indigenous knowledge system of central Cape York Peninsula), these are originally exogenous concepts. And the description of the accumulation of knowledge that follows, which speaks of '"diachronic" observations of the environment', matches Nader's (1996: 11) account of the 'scientific' accumulation of knowledge 'by verifying observation'. This is a world away from Thomson's (1933: 460) earlier observation that the 'traditional stock of knowledge' of the group involved in the oils project was held to have been 'invented' by their culture heroes or 'Stories' (see footnote 4). Yet, despite this apparent shift in local exegesis on the sources of knowledge, many local Aborigines continue to associate 'Stories' (including the 'rainbow serpent') with an 'immanent essence' that binds people, place, flora and fauna together. This essence is still understood as providing 'the object and sustaining force' of local understanding and action, inhering also in the 'moral soul' of each person and providing 'the binding force of the cosmos' (Wagner 1977: 404). Indeed, members of the group who describe their accumulation of knowledge through '"diachronic" observations' also commonly cite these 'Stories' and their 'law' as the foundation of the oils project.

This kind of hybridity makes it impossible to speak (analytically) of distinct knowledge systems in central Cape York Peninsula, despite many Aborigines desiring to understand the world in these terms (impelled, in part, by forms of imaginary which are part of the shared, intercultural field of regional development). Within the complex intercultural field in which they are now situated, a number of Aborigines – particularly older men and women – seek to navigate relationships with 'outsiders' by keeping much of their knowledge separate from their interactions with non-indigenes (knowledge about flying foxes and 'rainbows', for instance). Others (mostly younger men and women) actively seek to foreground similar knowledge in projects aimed at generating what they understand as opportunities both for cultural renewal and economic development. It remains to be seen what effects the resulting refiguration of local knowledge as 'indigenous knowledge', and even as 'local science', have on the local field of knowledge and practice. But it seems likely that – as has become increasingly apparent regarding the effects of 'globalisation' elsewhere – the result will be no more homogenous than 'global science'. Rather, particular localised formations of knowledge and practice will continue to proliferate. The hope remains that the Aboriginal people situated within this complex sociocultural field will find ways to effect more equitable exchanges of knowledge – as well as more satisfactory social, economic and political relationships – with the non-indigenes with whom they interact.

Acknowledgements

The Australian Research Council, the Leverhulme Trust, the Australian Institute of Aboriginal and Torres Strait Islander Studies, the Emslie Hornimann Fund, and the University of London Research Fund provided the financial support for the research on which this paper is based. I am deeply grateful for the ongoing support of the people of central Cape York Peninsula, without which this research would not be possible. In particular, I thank David Claudie, Willie Lawrence, Robert Nelson and MS (a deceased Kaanju woman from the Wenlock River) for their ongoing generosity and the insights which have arisen from our conversations. I am also indebted to Deirdre McKay, Bruce Rigsby, Paul Sillitoe and Jean-Christophe Verstraete for their comments on earlier drafts, which have aided me immensely in reworking this paper. I would also like to acknowledge Museum Victoria and Mrs Dorita Thomson for access to unpublished materials from the Thomson collection that I consulted in the process of drafting this chapter. The responsibility for any errors is mine.

Notes

1. Here and elsewhere in this paper, terms and phrases drawn from the local variety of Aboriginal English (see Nathan 1998; Rigsby 1998) are denoted by italics within inverted commas. Terms from (other) indigenous language varieties are simply italicised, with the language from which they are drawn identified in the text.

2. Anthropologists have described such '*medicine cut for woman*' (and similar substances and associated rituals employed by women) as 'love magic'. Other '*medicines*' were gathered by both men and women, and used to treat a variety of common ailments (hence the name), and for other purposes.

3. I have also discussed this meeting elsewhere (Smith 2005a). Indigenous knowledge pertaining to flying foxes – and their relationship to the rainbow serpent – is also discussed in McKnight (1975) and Wagner (2001: 132-35).

4. Across Cape York Peninsula, much of the knowledge that might be glossed as 'local science' is considered by many indigenous people as originating in the '*Story-time*' (elsewhere 'dreamtime'). During the *Story-time*, a set of figures whom Thomson calls 'culture heroes' or 'big men' (*yilamo* or *wulmpamo* in the Kuuku Ya'u language) are held to have 'invented the … culture and traditional stock of knowledge' (Thomson 1933: 460; also McConnel 1936; Rigsby 1999), which was then handed down to the region's contemporary Aboriginal population via their forebears. Aborigines now usually refer to these figures, as well as the accounts of their

actions, as '*Stories*'; the body of culture and local knowledge associated with these '*Stories*' is referred to as '*law*'. A core purpose of these accounts is to transmit knowledge about the local environment. Traditionally much of this knowledge was passed on to succeeding generations as part of the process of male initiation, often transmitted 'in the form of "sings" and legends' (Thomson 1933: 462). The local knowledge system in the Peninsula, like others of its kind, thus remains inseparable from local cosmology (Thomson 1933: 492). This knowledge system, like other knowledge systems across the globe, produces particular ordering effects, originally distinct to those that inhere in Western or 'global' science. However – as I discuss below – the interplay between 'Western' knowledge and 'local' knowledge in central Cape York Peninsula has led to a particular (albeit complex and 'internally' heterogeneous) contemporary knowledge system distinct from both the 'classical' indigenous knowledge system of the region, and the 'Western knowledge system' with which it has come to interact.

5.   A related point is made by Margolis, who suggests that the modes of human existence are 'constituted and reconstituted' by cultural forces 'in the same instant in which the "world" is constituted and reconstituted by our changing inquiries and interventions' (cited in Turnbull 2000: 3).

6.   See the papers in Hinkson and Smith (2005), especially Merlan (2005).

7.   As Nader (1996: 23) notes, '[local] knowledges are rapidly being overshadowed, replaced, and pushed aside by the introduction of a science and technology that assumes primacy. The recognition of lost knowledge, all but ubiquitously lost, is made all the more real by failed development projects.'

8.   As Sillitoe (1998: 227) notes, such assumptions of superiority can create 'a considerable barrier to development'.

9.   I acknowledge the critique made by Dove et al. (this volume) that debates around Traditional Environmental Knowledge have tended to reify both the difference between, and content of, 'indigenous' and 'Western' knowledge. But Wagner's analysis nonetheless remains pertinent in its illumination of often significant differences between the knowledge systems (and life-worlds more generally) apparent within modern, cosmopolitan sociocultural milieux and other fields of knowledge and practice. Moreover, such differences often remain vital to the character and outcomes of 'development' projects. Whilst we must take care not to exoticise and reify the 'Indigenous', we must also remain cognisant of the potentially radical (and often subtle) differences between the forms of cultural production that run through development practice and 'global science', and the local life-worlds with(in) which they engage.

10.   See also Descola and Pálsson (1996) for other anthropological engagements with the nature/culture dichotomy (and a discussion of previous anthropological positions on this issue, particularly in the introductory essay).

11.   The central Peninsula was settled by white miners and pastoralists in the late nineteenth century. Following a rapid process of dispossession, decimation and removal, the remnants of the region's indigenous population were incorporated into the region's new pastoral industry as an indentured labour force. Encapsulation led to considerable social and cultural disruption and a range of adaptive responses among Aborigines in the colonial era and in the postcolonial era of 'self-determination' that followed it. Both of these eras involved transformations of the local indigenous knowledge system, particularly through its 'intercultural' interplay with the originally exogenous knowledge system of the white settlers. I have discussed this history and its role in shaping contemporary indigenous life-worlds elsewhere (e.g., Smith 2003). Whilst it is difficult to say with any certainty what the nature of the local knowledge system was 'at the threshold of colonization' (Keen 2004), the work of early anthropologists in the region and elsewhere in northern Australia (much of which is summarised in Keen's monograph) – as well as aspects of the complex or 'hybrid' knowledge system of the contemporary region – suggest that the comparison made here with Wagner's 'indigenous Papuan' system is not inaccurate.

12. The term 'standing reserve' (Bestand in the original German – also glossed as 'commodity', 'raw-material' or 'resource') is associated with Heidegger (Young 1997: 176; also Smith forthcoming). In his discussion of Heidegger's work, Young (1997: 210-11, fn.31) notes the similarity between Heidegger's thinking of more 'reverential' relationships to *physis* (or 'nature'), and the 'guardianship' relations of Maori and other 'non- or pre-Western cultures'. The distinction between 'natural environment' and 'indigenous life-world' discussed here is closely related to the Heideggerian distinction between 'things' and 'equipment' (Haraway's 'resources'),

> a difference between an entity whose being is understood in terms of … [a] purely instrumental, power-directed set of categories, and one in which is understood in terms of a … multi-dimensional world, a world in which human beings can discover a 'home' [i.e., through what Ingold (2000) calls 'dwelling'] through discovering, simultaneously, their limitedness and their relatedness to everything that is (Young 1997: 203).

13. Weiner (2001) calls this relationship of dependence and elision 'nescience'.
14. Another, better documented project – the Traditional Knowledge Recording Project on 'Kuku Thaypan' (Awu Laya) homelands – has seen collaborative research undertaken by two senior Aboriginal men and researchers from James Cook University, aiming to 'provide sustainable solutions to current shortfalls in land management and water conservation strategies' (ISX 2005).
15. Whilst there is little doubt that many indigenous groups have a strong interest in protecting 'biodiversity', I think that there is a need for more critical attention to the use of the term 'sustainability' (which is linked to the culturally specific discourses about 'nature' and 'resources' discussed here) in intercultural contexts. This is particularly true when the context moves beyond 'environmental' questions to various forms of 'social' sustainability (e.g., that of governance arrangements) in milieux in which the collapse of various social arrangements has been a core aspect of 'social reproduction'. Such an enquiry is, however, beyond the scope of this paper.
16. Not to mention 'mainstream' Australian education, television, music, involvement in local corporations, waged employment etc.
17. See footnote 12.
18. See Smith (forthcoming).

References

Bicker, A., P. Sillitoe and J. Pottier (eds) 2004. *Development and Local Knowledge: New Approaches to Issues in Natural Resources Management, Conservation and Agriculture*. London: Routledge.

Brush, S.B. 1993. 'Indigenous Knowledge of Biological Resources and Intellectual Property Rights: The Role of anthropology', *American Anthropologist* 95(3): 653–86.

Descola, P. and G. Pálsson (eds) 1996. *Nature and Society: Anthropological Perspectives*. London: Routledge.

Franklin, S. 1995. 'Science as Culture, Cultures of Science', *Annual Review of Anthropology* 24: 163–84.

Foucault, M. 2004. '*Society Must Be Defended': Lectures at the Collège de France, 1975–76* (translated by D. Macey, edited by M. Bertani and A. Fontana). London: Penguin.

Haraway, D. 1988. 'Situated Knowledges: The Science Question in Feminism and the Privilege of Partial Perspective', *Feminist Studies* 14(3): 575–99.

Harding, S. 1994. 'Is Science Multicultural? Challenges, Resources, Opportunities, Uncertainties', *Configurations* 2(2): 301–30.

Harris, S.J. 1998. 'Introduction: Thinking Locally, Acting Globally', *Configurations* 6: 131–39.

Hinkson, M. and B.R. Smith (eds) 2005. 'Figuring the Intercultural in Aboriginal Australia', special issue of *Oceania* 75(3).

Ingold, T. 2000. *The Perception of the Environment: Essays in Livelihood, Dwelling and Skill.* London: Routledge.

ISX (Indigenous Stock Exchange). 'Traditional Elders Recognised by James Cook University'. www.isx.org.au/news [accessed 15 June 2005]

Keen, I. 2004. *Aboriginal Economy and Society: Australia at the Threshold of Colonisation.* South Melbourne: Oxford University Press.

Latour, B. 1993 [1991]. *We Have Never Been Modern* (translated by C. Porter). Cambridge, MA: Harvard University Press.

Lévi-Strauss, C. 1981. *The Naked Man* (translated by J. and D. Weightman). New York: Harper and Row.

McConnel, U. 1936. 'Totemic Hero Cults in Cape York Peninsula, North Queensland', *Oceania* 6: 452–47; 7: 69–105; 217–19.

McKnight, D. 1975. 'Men, Women, and Other Animals: Taboo and Purification among the Wik-mungkan', in *The Interpretation of Symbolism*, (ed.) R. Willis, pp. 77–97. London: Malaby Press.

Merlan, F. 2000. 'Representing the Rainbow: Aboriginal Culture in an Interconnected World', *Australian Aboriginal Studies* (2000/1&2): 20–26.

——— 2005. 'Explorations towards Intercultural Accounts of Socio-cultural Reproduction and Change', in M. Hinkson and B.R. Smith, (eds), 'Figuring the Intercultural in Aboriginal Australia', special issue of *Oceania* 75(3): 167–82.

Nader, L. 1996. 'Anthropological Inquiry into Boundaries, Power, and Knowledge', in *Naked Science: Anthropological Inquiry into Boundaries, Power, and Knowledge*, (ed.) L. Nader, pp. 1–25. New York: Routledge.

Nathan, D. 1998. 'Aboriginal English: a Cultural Study, by J.M. Arthur'. Review article. *Australian Aboriginal Studies* 1998(2): 87–89.

Pottier, J., A. Bicker and P. Sillitoe (eds) 2003. *Negotiating Local Knowledge: Power and Identity in Development.* London: Pluto Press.

Rigsby, B. 1998. 'Aboriginal English: A Cultural Study, by J.M. Arthur'. Review article. *Journal of the Royal Anthropological Institute* 4(4): 824–25.

——— 1999. 'Aboriginal People, Spirituality and the Traditional Ownership of Land', *International Journal of Social Economics* 26(7/8/9): 963–73.

Rowse, T. 2005. 'Indigenous Culture – the Very Idea'. Paper presented at the Human Geography Seminar Series, Research School of Social Sciences, The Australian National University, 2 May.

Scott, J.C. 1998. *Seeing Like a State: How Certain Schemes to Improve the Human Condition Have Failed.* New Haven, CT: Yale University Press.

Sillitoe, P. 1998. 'The Development of Indigenous Knowledge: a New Applied Anthropology', *Current Anthropology* 39(2): 223–52.

Sillitoe, P., A. Bicker and J. Pottier (eds) 2002. *Participating in Development: Approaches to Indigenous Knowledge.* London: Routledge.

Smith, B.R. 2003. '"Whither "Certainty"? Coexistence, Change and Some Repercussions of Native Title in Northern Queensland', *Anthropological Forum* 13(1): 27–48.

———— 2005a. '"We Got Our Own Management": Local Knowledge, Government and Development in Cape York Peninsula', *Australian Aboriginal Studies* (2005/2).

———— 2005b. 'Culture, Change and the Ambiguous Resonance of Tradition in Central Cape York Peninsula', in *The Power Of Knowledge, The Resonance of Tradition,* (eds), L. Taylor, G.K. Ward, G. Henderson, R. Davis and L.A. Wallis. Canberra: Aboriginal Studies Press.

Smith, B.R. forthcoming in 2006. 'Managing in a New Environment: Ecologies and Resources in Post-colonial North Queensland', *Cultural Studies Review* 12(1).

Thomson, D.F. 1933. 'The Hero Cult, Initiation and Totemism on Cape York'. *Journal of the Royal Anthropological Institute* 63(2): 453–537.

Tsing, A. 1999. 'Becoming a Tribal Elder and Other Green Development Fantasies', in *Transforming the Indonesian Uplands,* (ed.) T.M. Li, pp. 159–202. Amsterdam: Harwood.

Turnbull, D. 2000. *Masons, Tricksters and Cartographers: Comparative Studies in the Sociology of Scientific and Indigenous Knowledge.* Camberwell: Harwood Academic Publishers.

Wagner, R. 1977. 'Scientific and Indigenous Papuan Conceptualizations of the Innate: A Semiotic Critique of the Ecological Perspective', in *Subsistence and Survival: Rural Ecology in the Pacific,* (eds), T.P. Bayliss-Smith and R.G. Feachem, pp. 385–410. London: Academic Press.

———— 2001 *An Anthropology of the Subject: Holographic Worldview in New Guinea and Its Meaning and Significance for the World of Anthropology.* Berkeley: University of California Press.

Weiner, J. 2001. *Tree Leaf Talk: A Heideggerian Anthropology.* Oxford: Berg.

White, L., Jr. 1979. 'The Ecology of our Science', *Science* 80: 72–76.

Young, J. 1997. *Heidegger, Philosophy, Nazism.* Cambridge: Cambridge University Press.

5 On Knowing and Not Knowing: the Many Valuations of Piaroa Local Knowledge

Serena Heckler

In 1999, Slikkerveer wrote that local knowledge (LK) had developed 'almost parallel to Western "scientific", ... or "global" disciplinarity' (1999: 169), thereby imbuing local knowledge with the prestige often attributed to science. This claim that LK is rational and empirical is at the root of the participatory development approach championed by the likes of Chambers et al. (1989) and Warren et al. (1995). However, Agrawal (1995, 1999) points out that this claim is intensely value-laden and contributes to the assumption that a local system of knowledge can be accurately 'translated' into terms that are acceptable to scientists. In fact, Agrawal argues that this translation, which he calls 'scientization' (1999: 179), changes LK beyond recognition and makes it a tool of those who seek to exert control over marginalised peoples. He thereby suggests that the claim that local knowledge is somehow like science is, in itself, a dangerous misrepresentation. It can lead to an expectation on the part of scientists that 'knowledge' takes a form dictated by their particular disciplinary and theoretical paradigms.[1]

This paper tells of how I sought to set aside the preconceptions that had been inculcated in me through my training in the social and natural sciences to understand what the Piaroa of Amazonas State, Venezuela consider to be knowledge. In so doing, I uncovered five 'global' and 'local' knowledge paradigms, each of which are validated through their own designated authorities and each of which carry their own assumptions of value. I challenge the perception that there is a holism or objectivity to any of them. In fact, by placing the paradigms side by side, I demonstrate their dependence upon cultural validity and hence their tendency to select only certain types of information as valid, ignoring or discarding other types of information. Most importantly, I find, like Dove et al. (this volume), that the valuation of particular types of knowledge by global scientists

influences the choices local peoples make about what is desirable or even valid in
their own paradigms.

The Piaroa

Most Piaroa live in Estado Amazonas, Venezuela. Since the 1950s or 1960s, the
Piaroa have been moving from inaccessible highlands down to navigable river valleys, where they have more direct access to trade goods, biomedical health care and
schools (Mansutti 1988; Zent 1993). As of 2004, most Piaroa live in sedentary
and/or multiethnic communities along these river valleys, with a minority living in
small, semi-nomadic settlements that are rarely visited by outsiders. Because of the
difficulty of travel in Amazonas State, different communities and different regions
have been differently affected by various agents of development. For instance, some
communities have been thoroughly missionised by the New Tribes Mission, whereas in others shamanism maintains a strong influence. It is important to recognise
that the Piaroa, while only numbering 13,000 people, are heterogeneous.

Paradigm 1: LK and Ethnoscience

The background in which I was trained – biology, natural sciences and sustainable development – emphasised a cognitive ethnoscience approach to LK in
which nature is a constant that is essentially perceived the same way by all peoples, (e.g. Atran 1990; Berlin 1992). By elucidating a folk classificatory system,
which is primarily expressed through nomenclature, the minor variations in
what is an otherwise universal system of organising LK can be described. My
Ph.D. research project, based on sixteen months of fieldwork carried out
between 1996 and 1999, was largely designed to measure just that, specifically
focusing on Piaroa botanical classification. It was also a comparative study that
would measure how various aspects of globalisation affected Piaroa ethnobotanical knowledge in three communities in the Manapiare Valley region of
Amazonas State (Figure 5.1). Rather than enter into a critique of the scoring system or a discussion of the findings of this study, which I have published elsewhere (Heckler 2001, 2002), the concern here is to highlight the assumptions
embedded within the methods that I used and how that method was interpreted in the communities.

I used forest plot interviews, a method adapted from ecology that has become
central in ethnobiology (Phillips et al. 1994; Zent 1999). This method was first
used to determine what proportion of the forest is used sustainably and the
potential market value of such use—an economic argument against deforestation
(Peters et al. 1989). More recently it has been used to test hypotheses relating to
the impacts of various dependent variables on LK by carrying out structured
interviews with different people and comparing their 'knowledge' of the plot.

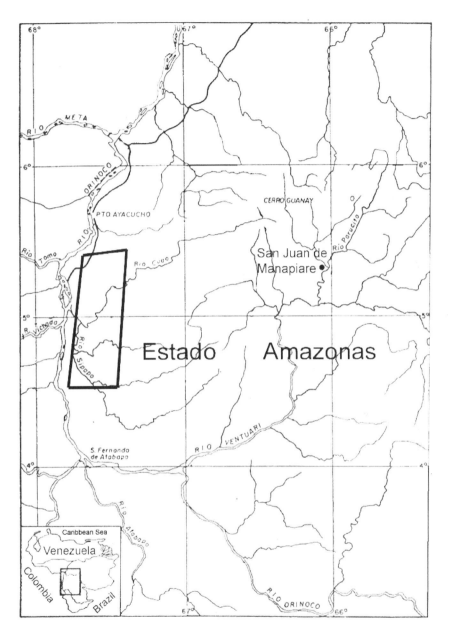

Figure 5.1 A map of Piaroa territory showing San Juan de Manapiare, the homebase of the research project. The box marks the approximate location of Overing's research, separated from the Manapiare Valley by highlands.

Because, as stated above, knowledge is assumed to be expressed through nomenclature, it can be measured by determining how many names and uses people can state.

I marked out plots in primary and secondary forest near three communities: one community was multiethnic; the second was relatively successful in the market economy; the third was more remote with less direct contact with the market economy, missionaries and other aspects of Venezuelan society. In each plot, I asked as many respondents as possible the names and uses of each plant. To deal with the problem of how to score names, which some researchers have dealt with through complicated quantitative methods (Phillips et al.1994; Phillips 1996), I supported the interviews with a variety of qualitative methods. I participated in activities involving the collection and use of wild plant resources; carried out informal interviews about other activities, particularly male ones such as basketry, hunting, house building and shamanic drug taking; and I cross-referenced anomalous answers with different respondents, often carrying sprigs of the plant around the community, asking people about the various names I had been given, discussing their origins and possible equivalency. In this way, I achieved a sufficient understanding of the species in my plots to identify correct or incorrect answers in 99 percent of responses. I scored responses for correctness, assigned a percentage to each interview and used these scores as a representative indicator of LK.

Even at this early stage, however, I was concerned that my method was overlooking important aspects of what the Piaroa considered to be knowledge. I was surprised when older women, whom I assumed would be knowledgeable about forest plants, claimed that they 'did not know' ('*tti cheruwa*') and so did not think they were the best people to be interviewed. Indeed, there was little correspondence between claims of knowledge and interview score. While it is possible that these women were simply showing humility, a quality highly valued by (older) Piaroa, other clues, including the ethnographic evidence below, led me to believe that my criteria for determining 'knowledge' differed from their criteria. I began to understand this disjunction more clearly when interviewing women about manioc cultivation. One woman, who cultivated five varieties of manioc in her gardens, claimed to 'not know' about manioc—pointing me to other women who 'did know'. Nevertheless, when she was shown manioc varieties from other women's gardens, she was able to name them and identify their tuber characteristics and for which preparations they were preferred. For her, being able to identify and name a species or variety represented only the most basic familiarity with the plant. There was some aspect of plant knowledge that plant names, hence my plot interviews, were not representing. I immediately suspected that they were referring to experience with the plant—gathering, growing or transforming it into usable products. This performance-based knowledge entails skills that cannot be verbalised, making it difficult to document, especially using the reporting-based methods favoured in participatory development and by ethnobiologists, including plot interviews.

Paradigm 2: LK as Performance

Throughout the 1990s, other authors were coming to this same conclusion and using it to critique the approach of participatory development practitioners and ethnobiologists. Sillitoe, whose writing on the subject is largely concerned with increasing integration between development workers and anthropologists (1998), argued that discrete data cannot be extracted from the broader performance of productive activities, thereby highlighting the impossibility of simply 'translating' LK into scientific frameworks that can then be channelled into development policy (2002; see also Marchand 2003). Richards dismissed efforts to elucidate systems of local knowledge as 'fallac(ies) of misplaced abstraction: the making of intellectual mysteries out of situations and activities whose practical import is obvious to all but the observer' (1993: 62). He argued that much LK is grounded in performance, practical, and highly responsive to changing circumstances (ibid.: 74–75). Ellen and Harris outlined a series of LK characteristics that include fluidity, its holistic, integrative and situated nature and its 'organization (as) essentially functional, denotative "know how" geared to practical response and performance' (2000: 4–5). From this perspective, LK is developed through trial and error, is highly practical and subject to change with changing needs and experiences. Those who adhere to this approach would say that the Piaroa acquire the bulk of their plant knowledge by engaging in activities, such as hunting and foraging, that utilise and generate that knowledge.

One of my most useful qualitative methods was participant observation while travelling through the forest with groups of people. Extended family groups took a day off their normal garden or hunting work to accompany me to my forest plots. People usually took the opportunity to gather fruits or insects that they knew to be ready for harvest in the vicinity, occasionally the men hunted and we often visited old gardens. By partaking in these activities children – and I – learned a myriad of details about forest ecology that could never have been verbalised, including the significance of the sound and smell of the forest, the humidity of the air, the quality of light filtering through the canopy. We also heard the adults speaking about these experiences, about what fruits and insects may be ready for harvest, and, crucially, using names to communicate this information. I realised that different family groups often used different names to refer to the various plant species. In fact my emphasis on finding the name for a plant occasionally stumped people, who then consulted with elders and shamans. Usually the shamans were able to provide a name, but sometimes, quite overtly, they made up a name on the spot, which was accepted by others and thereafter used to discuss the species. I also came across two systems of nomenclature: a formalised, ritual lexicon known only by elders and shamans; and an informal system, full of mnemonics and descriptive terms that are easily learned and used by children and young adults. The formal names were rarely elicited in the plot interviews, and shamans generally only used them when speaking to each other or when chanting during ritual. I will return to the importance of the ritual language later, but I had to accommodate its existence in scoring interview responses, in learning the language and in appreciating the fluidity of the bulk of the names elicited during the interviews. Although, as stated above, I was satisfied that I understood the sometimes complex and fluid nomenclature for the vast majority of the plants in my plots, my use of qualitative methods and my flexibility in accepting more than one name when scoring responses was criticised by other scientists as 'ad hoc', thereby implying that my findings were not valid. Admitting so much fluidity and so many variables into the scoring system obscured the replicability and transparency of the scientific method.

Nevertheless, based partially upon what I learned about the forest during these expeditions and partially upon my expectation that performance was the essence of LK, I was able to elucidate a well-rounded description of Piaroa botanical knowledge and correlate it to terminal indicators of that knowledge, i.e. plant names and uses. The logic was simple: the Piaroa go to the forest to obtain their livelihoods and in so doing, produce and use plant names in order to communicate their experiences. Not only did this make logical sense, but the remote community, where people spent the most time in the forest and depended most upon forest products, also scored significantly higher on the interviews (Heckler 2002: 541). The other two communities, one of which had little access to the forest due to localised deforestation, and the other which had ready access to manufactured goods, visited the forest much less frequently. In other words, my two methods cross-checked, demonstrating that the two methods, when applied carefully, were not incommensurable. I thus validated the use of plot interviews as a shortcut for evaluating LK, even if, in modifying the method so that it reflected a system much more complex than allowed for by the underlying theory, it was no longer accepted as valid by those who first developed the method. I also demonstrated a means by which the Piaroa empirically learned from their environment and the outcome of that learning, subsistence skills, a type of 'local science'.

Disconcertingly, however, I had not solved my initial problem. Piaroa claims of knowledge did not correlate with the findings of my study. In the most remote community, claims of ignorance correlated with interview score only for children. Why did elderly women, who had more experience than almost anyone else, claim to 'not know'? Why did some of the most dedicated and successful hunters and fishers tell me that they 'didn't know'? To answer these questions, I delved more deeply into the literature, making connections that had not previously been apparent, and carrying out more focused interviews. I came upon a uniquely Piaroa conception of knowledge that ran contradictory to the paradigms that I had described in my own research.

Paradigm 3: Ta'kwąyą – Piaroa Local Knowledge

Of the early Piaroa ethnographic studies, the work of Overing stands out in its thoroughness and accuracy. Her observations on the cosmology and social organisation of the Piaroa have been repeatedly corroborated (Zent 1992: 6–7; Oldham 1996, 1997; Heckler 2004). In the late 1960s, Overing and her partner, Mike Kaplan carried out fieldwork with the Piaroa of the Paria, Cuao and Sipapo Rivers (see Figure 5.1) and wrote about their social development (Kaplan 1975; Overing 1986, 1989a, 1989b, 1993; Overing and Kaplan 1988). In one article (1989b), she discusses the way in which Piaroa children acquire the knowledge that enables them to become properly domesticated Piaroa adults. Embedded within a much broader cosmology, Overing describes the concept of *ta'kwąyą*, which she translates as 'the knowledge of, and the capabilities for using, the customs of one's own people, including its language, its social rules, its processing of food, its ritual' (1989b: 175), in other words, something very like LK. *Ta'kwąyą* is acquired through shamanism, specifically through a series of rituals, called *maripa te'au*, that shamans administer to children as they grow. During the maripa te'au the shaman 'chants to the gods to give the child its ... knowledge and capabilities for gardening, acquiring food from the forest and so on' (1989b: 180). The knowledge necessary for carrying out subsistence activities, rather than being learned through doing, is held to be acquired from the crystal boxes of the gods (Overing 1989b: 186; 1993: 198) where the potentially dangerous knowledge of cultural skills is kept in the form of crystal beads.

She describes how, as people grow older, they have a choice of whether to continue to receive lessons from the shamans throughout their early adulthood, thus gaining knowledge, power and eventually becoming shamans, or carry out their normal daily tasks with the minimum knowledge required and remain 'ignorant'. Overing gives an example of how these different roles are perceived: she writes of a hunter who brought back a great deal of meat but was said to be a poor hunter because he did not 'know':

> The practical hunter had not gone beyond his boyhood (learning ceremony) for the hunt and the capabilities given to him by the gods during that ceremony allowed him the skills necessary for killing game in the forest ... (but) (magical) capabilities also give the (shaman) the knowledge to transform forces from other worlds and it is through his powers for transformation that he makes fertile the forest and supplies it with game for the benefit of the practical hunter. (1989b: 187)

What it means to be a knowledgeable hunter is not to bring the meat back to the camp, but to call it forth from its home with the gods and to make it edible.

Similarly, Overing mentions old women who are very skilled at particular tasks and highly fluent in the ritual language of the elders, but still profess their ignorance, their vulnerability and their ultimate dependence upon men (1989b: 188). According to Overing, underlying the experiential knowledge emphasised in performance-based studies of LK is a theoretical knowledge that enables a well-educated Piaroa to live successfully within his or her environment. Most importantly, the experiential knowledge of how to hunt, how to plant, how to weave a basket and which plant species to utilise in these tasks – that knowledge which is highly fluid and practical – is not valuable. On the other hand, the knowledge that is passed down from shaman to apprentice and which is learned in highly formalised rituals – codified and arcane knowledge – is highly valued by the Piaroa. According to this telling of it, neither the classificatory knowledge that my quantitative research measured nor the experiential knowledge sought and elucidated through my qualitative research reflect Piaroa epistemology. The implications of this idea for proponents of local knowledge are significant. The experientialists are highly critical of attempts to describe formal knowledge systems, arguing that they are scientific abstractions (Richards 1993), and yet, by Overing's telling of it, the very basis of the knowledge that the Piaroa hold to be valuable is a formalised, highly elaborated paradigm. The cognitivists are intent on a very particular type of knowledge being the representation of humans' evolved cognitive structure and, in fact, the *ta'kwąyą* paradigm holds that nomenclature, or rather the highly formalised language of the elders that the shamans use to communicate with the gods, to call forth game animals and to bring the beads of knowledge back to individuals, is highly valued. However, these are not the names that I normally elicited during my plot interviews, nor is the Piaroa understanding of the significance of these names acceptable to the cognitivist paradigm. Not only had the quantitative methods excluded the possibility of eliciting a paradigm that differed substantially from the scientific model, but the experiential approach had encouraged me to consider only the praxis, not the abstract model underlying it.

However, as always, there are caveats. Firstly, Overing, working within the anthropological paradigm of the day, privileged a structured, internally coherent cosmology over subsistence activities. In fact, the Piaroa value both *ta'kwąyą* and practical skill. Certainly, the practical hunter may not have the knowledge to call forth the game animal or to transform it from dangerous flesh to something that is safely edible, but his ability to bring home meat regularly is acknowledged by his co-residents. Likewise, an 'ignorant' woman's ability to cultivate and prepare plentiful manioc beer or bread is esteemed, though more as an expression of moral rectitude than as a highly developed skill (Heckler 2004). Most importantly, however, my informants never spontaneously volunteered information about *ta'kwąyą* or related concepts, prompting me to incorporate questions about shamanic learning rituals, knowledge and *ta'kwąyą* into my informal interviews. I found an almost complete lack of comprehension of the topic specifically and only a vague idea of the importance of shamanic ritual for enculturation and childhood development. Not one person I spoke to was able to answer the direct question: '¿Qué significa la palabra '*ta'kwąyą*'?' (What does the word *ta'kwąyą* mean?) and, although most understood the term '*maripa te'au*', only a few had undergone the ritual.

There are several possible reasons for this. Firstly, Overing may have got it wrong. However, given the accuracy of most of her material, I do not have the evidence necessary to discount a concept that she has described in great detail. Secondly, most Piaroa, like people everywhere, do not reflect upon the cosmological underpinnings of what they know. It is a specialised knowledge belonging to shamans and Piaroa shamans are notoriously secretive (Oldham 1996). Given that my research was not specifically concerned with shamans, I did not make the effort necessary to develop relationships with them, as Overing did. Even when shamans chose to discuss plant names with me, they did not explain how they knew these names or where their knowledge came from. Finally, and most importantly, my interviews were carried out some thirty years after Overing's interviews – during which time the lived experience of the Piaroa changed a great deal.

Paradigm 4: Evangelism, Science and Politics

In 2004, I returned to the Manapiare Valley to ask questions about ta'kwąyą and the contradictions between my own findings and Overing's research. Many of the Piaroa to whom I spoke made it clear that their conversion to Evangelism by the North American New Tribes Mission (NTM) had greatly undermined shamanic authority. They told me that the NTM explicitly forbids shamanic practices and they believed that shamanism is evil, superstitious and only believed by the 'ignorant'. My informants told me that shamanism is 'holding them back' and that they need to 'leave it behind them'. Some even refused to speak about it, saying that it was a 'bad' subject. In other words, they are not simply forgetting shamanic cosmology, rather they are actively rejecting it, blaming it for the power imbalances between indigenous and non-indigenous people. A few shamans still exist, but they are no longer central authorities in the community. Rather than being the patriarch around whom a community forms, they often travel from community to community visiting kin. Rather than singing every night to maintain the productivity of the forest and maintain amicable relations with forest spirits, they usually only carry out healing rituals. Rather than dictating the context in which knowledge is obtained and maintained, they have become marginalised figures, to be treated with caution and ambivalence.

Evangelism has eroded belief in shamanism, but is not able to entirely fill the void it has left. Whereas shamans are concerned with the spiritual, intellectual, social and physical development of their community members, Evangelism is primarily concerned only with what it considers to be the spiritual, thereby failing to offer a knowledge paradigm to replace the shamanic ta'kwąyą. Instead it comes bundled with a variety of exogenous phenomena, including market economies, national and state politics and the positivist philosophy of visiting scientists, who take on some of the authority no longer vested in shamans. As the Piaroa were converting to Evangelism, they were finding a host of new 'authorities' who, they hoped, would show them how to make money and to be successful in this new society.

It was not until the late 1980s that anthropologists established long-term relationships with the residents of the Manapiare Valley. One of these anthropologists, Stanford Zent, has collected extensive data on plant knowledge in the region, while another researcher carried out a limited survey of medicinal plant use. By the time I first visited the area in 1996, the Piaroa fully expected scientists to be interested in plants, primarily plant names and use. Indeed, some of them told me that I was 'doing it wrong' when my methods differed from Zent's. My research only corroborated their suspicions that common plant names and uses – considered superficial by the shamans – had some value in the outside world. This local perception has been strengthened by national, political rhetoric condemning biopiracy and 'gringo imperialism', a rhetoric that some Piaroa have adopted as they become increasingly involved in national and regional politics.

Throughout my fieldwork, Piaroa leaders attended political meetings and took stances on exogenous concepts, including LK (regionally known as *conocimiento tradicional*). Much of the information that they have received on these topics has been filtered through an indigenous organization that serves as a central advocate for indigenous rights in Puerto Ayacucho, the Regional Organisation of Indigenous Peoples, Amazonas (ORPIA). In November 2002, it held a conference at Tobogán de la Selva to inform indigenous representatives of a case where medicinal plant knowledge had been stored in a computer database with a promise that the information would be 'returned to the community'. The database was subsequently removed from the public domain in anticipation of sale at a vast profit, none of which is expected to accrue to the Piaroa people. They were also informed of several other high-profile scandals in which researchers have illegally removed botanical specimens from the area and/or recorded and attempted to profit from ethnomedicinal knowledge. The outcome of this conference, at least according to ORPIA, was a request for a moratorium on all LK research until clear guidelines are established. As a result, the government's Office of Indigenous Affairs is extremely reluctant to grant permit applications to projects that even mention the words '*conocimiento tradicional*'.

However, according to most of the Piaroa to whom I spoke, the moratorium is only applicable to 'research that they do not want', that is, 'bad research'. When I asked for clarification, they said that research that can demonstrably accrue material benefits to the local communities is 'good', while research that does not do so is 'bad'. While attempts have clearly been made to profit from LK without equitable benefit-sharing agreements, some projects, including my own, that have gone to great lengths to avoid the commodification of LK are widely suspected of being 'bad'. It is assumed that if the profits are not seen by local communities, they must be accruing elsewhere, a viewpoint that is strengthened by the promises of economic return liberally and naïvely offered by some bioprospectors. Despite years of concerted effort to steer clear of the bioprospecting debate,[2] the plot interviews that I undertook back in 1999 have been held up as an example of materially beneficial research, and the names and uses of plants that I recorded have come to be seen as one of the most valued kinds of knowledge. The fact that I have not set up legal agreements stipulating the distribution of the profits that my research is presumed to be generating,[3] that I have not helped the Piaroa to take out patents (locally understood as 'a grant for life' [Sp. *beca para la vida*]) has laid me open to suspicions of biopiracy. If this were the end of the story, I would have been refused permission to enter any of the research communities and I would have given up in despair. However, another aspect of national political discourse has begun to shift the focus on LK away from shallow materialism and onto its potential significance for indigenous cultural, social and land rights.

Paradigm 5: Indigenous Rights and a New Tradition

Hugo Chavez's government, elected in 1998, has brought indigenous rights to the forefront of the national political agenda. They were incorporated into the federal constitution in 1999 (Title III, Chapter VIII), a national commission of indigenous peoples has been created that liases with the central administration and a decree has been passed ensuring the demarcation of indigenous territories. To take full advantage of these opportunities, each indigenous group must demonstrate their ethnic distinctiveness, they must create formal political structures that the government will recognise and they must prove past or present use of land and resources. The resulting reorganisation in indigenous communities throughout Amazonia has changed the communities in a variety of ways, two of which are significant to this discussion. The first is that many people to whom I spoke were interested in documenting and archiving their LK, in part because it is required by the governmental procedures set up for distributing resources earmarked for indigenous peoples. But they also, influenced by the government's emphasis on 'traditional culture', believe that they must 'rescue their traditional knowledge' and 'preserve it for future generations'. I will return to this point in a moment, but first I would like to consider the second recent development, the creation of 'councils of elders'.

None of the communities that I visited in the 1990s had 'councils of elders', nor did anyone mention them to me. By 2004, however, they had become widespread. The idea seems to have been borrowed from the international indigenous rights movement, whence it made its way into governmental and NGO procedures. The government required that decisions be ratified by a 'council of elders' and meetings, such as Tobogán de la Selva, set aside space in the agenda for councils of elders to meet and make statements. As a result, local communities have had to form such councils, which include shamans.[4] One result of these bodies is that shamans begin to be reintegrated into community governance and, crucially for Amazonians (see Clastres 1989), are given a forum in which to make public statements.

At Tobogán de la Selva, according to the official transcript of the meeting, the shamans expressed concern and some bemusement that local knowledge should be considered extractable from its context and they took offence that it should be considered a marketable product. Instead, they claim that they are the 'scientists' that 'understand how to educate, how to visualise, how to transmit to generation(s) that really will defend this knowledge' (Juan Antonio Bolívar, Piaroa shaman). Thus, the shamans are reasserting their authority to define what knowledge is, to whom it should be transmitted and in what form.

This is not to say that they have regained their positions as central figures in Piaroa society nor are they allowed to set the agenda in political meetings. Rather they are expected to offer the 'traditional' perspective on controversial issues. This perspective is translated by younger, politically active Piaroa, many of whom are evangelicals, to whom only certain types of 'traditionality' would be acceptable. The statements are then transcribed by Venezuelan or mestizo employees of the indigenous organisations before being distributed back to the indigenous communities to be pored over and back-translated to largely monolingual community members. By the time the statements reach the community, they reflect what the politicians expect or want the shamans to say rather than what the shamans may have intended. Instead of encouraging a return to shamanic authority as described by Overing, this process is creating a new paradigm that blends a variety of influences, including indigenous rights politics, Christian concepts of spirituality, and international environmentalism with Piaroa ideas about how to survive in a rapidly shifting political and social climate.

The local concern with 'rescuing traditional culture' and the representation of shamans' views by young, indigenous politicians troubled me at first – I had visions of museum cultures being manipulated and misrepresented by slick, Piaroa politicos – but discussions with ordinary Piaroa farmers, hunters and labourers helped me to see another aspect of the issue.

One thirty-year-old Evangelical man told me that the point of the *maripa te'au* is that the shaman will take 'Piaroa thoughts' (Sp. *pensamientos*) from the place of the gods and put them into socially developing children. This man did not consider the thoughts nor the *maripa te'au* to be important, but rather that the ritual places the child's development at the centre of the community's activity. When I asked a twenty-year-old Piaroa man, who believed that shamanism was 'old-fashioned', what most defined Piaroa-ness, he replied, 'you must be able to communicate with other Piaroa'. This deceptively simple statement is significant given the importance of 'speaking well' and 'living well' that has been noted throughout Amazonia (Gow 2001; Belaunde 2001). Or, to borrow from the indigenous rights movement that is so influential in this paradigm:

> 'What is 'traditional' about traditional knowledge is not its antiquity, but the way it is acquired and used. In other words, the social process of learning and sharing knowledge, which is unique to each indigenous culture, lies at the very heart of its 'traditionality'. (The Four Directions Council [1996] in Posey 1999: 4)

From this perspective, it is not what is known that is important, but the context in which learning and teaching occur. From the point of view of these ordinary Piaroa, the new emphasis on 'traditional culture' does not reflect a desire to objectify their past, but to ensure a continuation of kinship-based sociality, a central concern for peoples throughout Amazonia (Overing and Passes 2000; McCallum 2001; Johnson 2003).

Conclusion

My basic point is simple: what is considered to be knowledge depends upon what is considered to be valuable. Science, with its mandate to further knowledge, often ignores or devalues types of knowledge that are not in line with its latest paradigm (Kuhn 1970). This paper has demonstrated how scientists who ostensibly engage with the same issue may seek out, collect and analyse completely different types of information. In this case, seeking to elucidate 'local knowledge' may mean eliciting plant names and uses, describing skills and experiences or describing abstract cosmological models, depending on the theoretical stance of the scientist.

What makes this paper more than a methodological critique, however, is the fact that science holds a unique authority to decide what is valuable, not only within its own circles, but also in wider society. The Piaroa, who until recently turned to shamans to acquire the knowledge that they needed to live well, have increasingly turned to other authorities to tell them what is and is not knowledge, what is and is not ignorance and what aspects of their LK they should maintain and what should be discarded. Scientific valuations, interpreted by local peoples through the perspectives of other exogenous authorities, have helped to change the types of knowledge that the Piaroa privilege, hence their very epistemology.

I certainly do not wish to imply that one paradigm is 'right' and another 'wrong'. Nor do I suggest that scientists should not work within particular disciplinary perspectives. Indeed philosophers of science agree that it is impossible to pursue knowledge without an interpretive horizon (Gadamer), or, as Kuhn would put it, a paradigm (1970: 16). However, it is essential that people from different perspectives, both local and scientific, meet and interact with each other. Within global science, within local societies and between global science and local societies, people negotiate different perspectives and, hopefully, further knowledge through acknowledging these differences. I hope that, by bringing together paradigms that generally operate separately from each other, I have offered a dialogue that can improve our understanding of the Piaroa and of those who study them.

Acknowledgements

I thank the Piaroa with whom I have spent many hours discussing these issues. I only hope that I have fairly represented their concerns. I also thank Paul Sillitoe for his helpful comments on earlier drafts of this paper. This research was made possible by two grants from the ESRC (T026271416 and RES-000-22-0689), from grants by the Fulbright Foundation, the Garden Club of America, the Cornell Research Training Group in Conservation and Sustainable Development and the L.H. Bailey Hortorium.

Notes

1. Here I am borrowing from Kuhn (1970), who describes paradigms as 'laws, theories, applications and instrumentations, etc. which provide models from which spring coherent traditions' (1970: 10).
2. I have intentionally withheld medicinal plant data that are not already within the public domain, I have not focused on nor taken herbarium specimens of medicinal plants and I have intentionally kept my research independent of would-be collaborators who are interested in bioprospecting.
3. I have, in all communities and at all stages of my research, drawn up consent agreements that include material contributions to the community. An important stipulation of these agreements is the return of study results to the communities, one of the primary aims of the 2004 fieldwork. The agreements and the monetary benefit accompanying them were not acknowledged by those condemning my research.
4. This formalised, hierarchical structure is at odds with the fluid and non-hierarchical political structure typical throughout this region of Amazonia (Riviére 1984). For a much more thorough analysis of the effects of national political structures on the Piaroa see Oldham (1996).

References

Agrawal, A. 1995. 'Dismantling the Divide between Indigenous and Scientific Knowledge', *Development and Change* 26: 413–39.
———— 1999. 'Ethnoscience, 'LK' and Conservation: on Power and Indigenous Knowledge in *Cultural and Spiritual Values of Biodiversity*, (ed.) D.A. Posey, pp. 177–84. London: Intermediate Technology Productions.
Atran, S. 1990. *Cognitive Foundations of Natural History*. Cambridge: Cambridge University Press.
Belaunde, L. 2001. *Viviendo bien: género y fertilidad entre los Airo-Pai de la Amazonía Peruana*. Lima: CAAAP.
Berlin, B. 1992. *Ethnobiological Classification: Principles of Categorization of Plants and Animals in Traditional Societies*. Princetown NJ: Princeton University Press.
Bicker, A., P. Sillitoe and J. Pottier (ed) 2004. *Development and Local Knowledge: New Approaches to Issues in Natural Resources Management, Conservation and Agriculture*. London: Routledge.

Brokensha, D., D. Warren and O. Werner (ed) 1980. *Indigenous Knowledge Systems and Development*. Lanham, MD: University Press of America.

Chambers, R., R. Pacey and L. Thrupp (ed) 1989. *Farmer First: Farmer Innovation and Agricultural Research*. London: Intermediate Technology Publications.

Clastres, Pierre 1989. *Society against the State: Essays in Political Anthropology*. New York: Zone Books.

Ellen, R.F. and H. Harris 2000. '*Introduction*', in Indigenous Knowledge and Its Transformations, ed, R.F. Ellen, P. Parkes and A. Bicker. Amsterdam: Harwood Academic Publishers.

Ellen, R.F., P. Parkes and A. Bicker (ed) 2000. *Indigenous Knowledge and Its Transformations*. Amsterdam: Harwood Academic Publishers.

Gow, P. 2001. *An Amazonian Myth and Its History*. Oxford: Oxford University Press.

Heckler, S. 2001. 'The Ethnobotany of the Piaroa: Analysis of an Amazonian People in Transition'. Unpublished Ph.D. thesis, Cornell University, Ithaca, NY.

——— 2002. 'Traditional Knowledge and Gender of the Piaroa', in *Ethnobiology and Biocultural Diversity*, ed, J.R. Stepp, F. Wyndham and R. Zarger. Proceedings of the VIIth International Congress of Ethnobiology. Athens, Georgia: International Society of Ethnobiology.

——— 2004. 'Tedium and Creativity: the Valorisation of Manioc Cultivation by Piaroa Women', *Journal of the Royal Anthropological Institute* 10: 241–59.

Johnson, A.W. 2003. *Families of the Forest: the Matsigenka Indians of the Peruvian Amazon*. Berkeley, CA: University of California Press.

Kaplan, J. 1975. *The Piaroa, a People of the Orinoco Basin: a Study in Kinship and Marriage*. Oxford: Clarendon Press.

Kuhn, T. 1970. *The Structure of Scientific Revolutions*, 2rd edition. Chicago and London: The University of Chicago Press.

Mansutti Rodriguez, A. 1988. 'Pueblos, comunidades y fondos: los patrones de asentamiento Uwotjuja', *Antropológica* 69: 3–36.

Marchand, T.H.J. 2003. 'A Possible Explanation for the Lack of Explanation; Or, "Why the Master Builder Can't Explain What he Knows": Introducing Informational Atomism against a "Definitional" Definition of Concepts', in *Negotiating Local Knowledge: Power and Identity in Development*, ed, J. Pottier, A. Bicker and P. Sillitoe. London: Pluto Press.

McCallum, C. 2001. *Gender and Sociality in Amazonia: How Real People Are Made*. Oxford: Berg.

Oldham, P. 1996. 'The Impacts of Development and Indigenous Responses among the Piaroa of the Venezuelan Amazon', Ph.D. thesis, London School of Economics, University of London.

————— 1997. 'Cosmología, Shamanismo y Práctica Medicinal entre los Wothïha (Piaroa)', in J. Chiappino and C. Alés, ed, *Del Microscopio a la Maraca*. Caracas: Ex Libris. pp. 225–49.

Overing, J. 1986. 'Men Control Women? The "Catch 22" in the Analysis of Gender', *International Journal of Moral and Social Studies* 1(2): 135–56.

————— 1989a. 'The Aesthetics of Production: the Sense of Community among the Cubeo and Piaroa', *Dialectical Anthropology* 14: 159–75.

————— 1989b. 'Personal Autonomy and the Domestication of the Self in Piaroa Society', in *Acquiring Culture: Cross Cultural Studies in Child Development* ed, G. Jahoda and I.M. Lewis. London: Croom Helm.

————— 1993. 'Death and the Loss of Civilized Predation among the Piaroa of the Orinoco Basin', *L'Homme* XXXIII (2–4): 191–211.

Overing, J. and M.R. Kaplan 1988. 'Los Wothuha (Piaroa)', in *Los Aborigenes de Venezuela: Etnología Contemporánea II,* (ed.) Walter Koppens pp. 307–411. Caracas: Fundación La Salle de Ciencias Naturales, Instituto Caribe de Antroplogoía y Sociología.

Overing, J. and A. Passes 2000. *The Anthropology of Love and Anger: the Aesthetics of Conviviality in Native Amazonia*. London: Routledge.

Peters, C., A. Gentry and R. Mendelsohn 1989. 'Valuation of an Amazonian Rainforest', *Nature* 339: 655–56.

Phillips, O. 1996. 'Some Quantitative Methods for Analyzing Ethnobotanical Knowledge', in *Selected Guidelines for Ethnobotanical Research: a Field Manual,* (ed.) M. Alexiades. New York: New York Botanical Gardens.

Phillips, O., A.H. Gentry, C. Reynel, P. Wilkin and C. Gálvez-Durnd 1994. 'Quantitative Ethnobotany and Amazonian Conservation', *Conservation Biology* 8: 225–48.

Posey, D. 1999. *Cultural and Spiritual Values of Biodiversity: A Complementary Contribution to the Global Biodiversity Assessment.* London: Intermediate Technology Publications.

Richards, P. 1993. 'Cultivation: Knowledge or Performance?', in *An Anthropological Critique of Development: the Growth of Ignorance* (ed.) M. Hobart pp. 61–78. London: Routledge.

Riviére, P. 1984. *Individual and Society in Guiana: a Comparative Study of Amerindian Social Organization.* Cambridge: Cambridge University Press.

Sillitoe, P. 1998. 'The Development of Indigenous Knowledge: a New Applied Anthropology', *Current Anthropology* 39(2): 223–52.

————— 2002. 'Participant Observation to Participatory Development: Making Anthropology Work', in *Participating in Development: Approaches to Indigenous Knowledge,* ed, P. Sillitoe, A. Bicker and J. Pottier pp. 1–23. London: Routledge.

Slikkerveer, L.J. 1999. 'Ethnoscience, 'LK' and Its Application to Conservation', in *Cultural and Spiritual Values of Biodiveristy,* (ed.) D.A. Posey pp. 169–259. London: Intermediate Technology Publications.

Warren, D.M., L.J. Slikkerveer and D. Brokensha (ed) 1995. *The Cultural Dimensions of Development: Indigenous Knowledge Systems.* London: Intermediate Technology Publications.

Zent, Stanford 1992. '*Historical and Ethnographic Ecology of the Upper Cuao River Wōthihā: Clues for an Interpretation of Native Guainese Social Organization*'. Unpublished D.Phil. dissertation, Columbia University.

———— 1993. 'Donde No Hay Médico: las consecuencias culturales y demográficas de la distribución desigual de los servicios médicos modernos entre los Piaroa', *Antropológica* 79: 41–84.

———— 1999. 'Acculturation and Ethnobotanical Knowledge Loss among the Piaroa of Venezuela: Demonstration of a Quantitative Method for the Empirical Study of Traditional Ecological Knowledge Change', in *On Biocultural Diversity: Linking Language, Knowledge, and the Environment,* (ed.) L. Maffi. Washington: Smithsonian Institute Press.

6 The *Ashkui* Project: Linking Western Science and Innu Environmental Knowledge in Creating a Sustainable Environment

Trudy Sable with Geoff Howell,[1] Dave Wilson[2] and Peter Penashue

Special thanks to Julia Sable

Intuitively, it seems impossible to talk of development without the inclusion and consultation of the people whose lives will be affected, aboriginal or not. Furthermore, aboriginal populations have a wealth of knowledge accumulated over centuries of living in their regions that can enhance government efforts to protect the environment. The obligation to protect and include aboriginal communities as part of environmental conservation and sustainable development initiatives is enshrined in several international declarations, including the Rio Declaration and the Convention on Biological Diversity (CBD), as well as Canadian legislation and declarations (Canadian Environmental Protection Act 1999). However, the relationship between scientific research and community involvement and capacity building is still a much debated issue. This debate is not just among scientists in regard to the validity of aboriginal peoples' environmental knowledge, but among aboriginal peoples themselves who are asking how scientific research can serve their needs given its long association with colonisation and industrial development projects (Marzano, this volume). The legacy of colonisation, and the subsequent institutions and development projects that have accompanied it, has often served to dislocate aboriginal peoples from their lands and discount their environmental knowledge as inferior (Bieder 1986). As a result, an atmosphere of distrust exists among aboriginal peoples toward social and natural scientists. This paper will examine a collaborative project called the *Ashkui* Project (a.k.a. the Labrador Project). The *Ashkui* Project is an example of how scientists and aboriginal communities can work collaboratively to redress this legacy, and build trust, with the ultimate goal of building capacity among all stakeholders around a common vision.

Since 1997, the federal department of Environment Canada has collaborated with the Innu Nation (First Nation) of Labrador and social scientists from the Gorsebrook Research Institute (GRI) at Saint Mary's University in Halifax, Nova Scotia to incorporate social sciences and community involvement into environmental research. The initial goal was to develop comprehensive baseline ecological data of the Labrador landscape from both Innu and Western scientific perspectives. To provide an initial focal point for research, consultations were held with members of the Innu community to identify an aspect of the landscape that was deemed culturally significant and distinct. The landscape feature the Innu chose is *ashkui*, giving rise to what was referred to as the *Ashkui* Project in Labrador, Canada.

Ashkui have been defined by Innu elders (*tshishennuat*) as early openings in the water. The importance of *ashkui* comes in the springtime when an abundance of waterfowl, fish and animals are attracted to the newly opened water. Innu families set up camp near these sites and stay for weeks and months at a time to exploit and celebrate the rich and varied resources of the land and water. These sites are still of importance to Innu for their livelihood and psychological well-being. Conveniently, *ashkui* were also the home of the Harlequin duck, an endangered species that Environment Canada was mandated to study. Conducting research at these sites would allow for Environment Canada's continued focus on waterfowl inventories and hydrometric research.

GRI social scientists, working closely with *tshishennuat* and Innu co-researchers/Guardians, have documented some of the complex knowledge of those Innu who grew up on the land, and whose lives were inextricably linked to the *ashkui* sites. Based on this research and the identification of a number of *ashkui* sites that were in current use or had been regularly used in the past by Innu families, Environment Canada scientists in the spring of 1999 set up a 325 km *ashkui* research network comprising fifteen stations. These sites were used to sample and analyse three freshwater systems – clear, coloured and brackish. Since 2001 the project has expanded to include the Innu Environmental Guardians Program, an initiative to assist in training members of the Innu community in ecosystem management in preparation for self-government. The project has a commitment to incorporate Indigenous Ecological Knowledge (IEK) and to work with the Innu Nation in the overall project design and structure.

We will discuss some of the crucial aspects of this research effort in documenting and working with the different perspectives of the landscape, Innu and Western scientific, and setting up cross-cultural dialogues. In particular, we will focus on the concomitant capacity building and shifts in thinking that, by necessity, had to occur among all stakeholders in order for the research to serve both the needs of the community and the mandate of Environment Canada to monitor and protect the environment.

Background and Innu Perspective

The *tshishennuat* commonly speak of their ancestral lands, *Nitassinan*, as medicine.[3] They often complain of the mental and physical sickness that has afflicted them since government settlement programmes in the 1960s isolated them from their traditional lands. The food they eat is no longer 'medicine' because it does not come from the land. Even the animals they used to hunt are confused and sick, and do not always follow predictable migration routes. Furthermore, the *tshishennuat* say they can no longer communicate with their youth, who are being educated in unfamiliar ways and in a foreign language. Within their communities, there is concern that the Innu youth no longer have the same relationship to the land as the *tshishennuat*, and are losing the language of the land. This is worrying because *tshishennuat* are the holders of a complex and specialised knowledge of the environment, which has been passed down orally through centuries. They are the speakers and holders of the language of the land and carry the Innu 'library' of knowledge in their heads. Because of their knowledge and relationship to the land, they are the best to advise and define the needs of any environmental development. This library will be lost with their passing.

The loss of such an extensive body of knowledge has become a global issue that is gaining widespread advocacy in the fields of conservation and bio-diversity as the link between cultural diversity and biodiversity becomes evident (Davis 2002). Indigenous peoples, such as the Innu, intimately understand the fragility of their local ecosystems, whether it be the boreal forest of Labrador or the Himalayan mountains of Bhutan. Their language, songs, dances and legends reflect and express its many aspects, creative and destructive. With the loss of more than 3,000 languages within the last fifty years, the languages that have articulated this environmental knowledge is lost, and the language of science is increasingly being used to replace it (Davis 2002; Maranzo, this volume).

Metaphorically speaking, finding the medicine again has been the challenge of the last eight years of collaboration with the Innu. Due to the many pressures facing their communities from development projects, the Innu have hired scientists, anthropologists and a variety of consultants to help in their quest for self-determination. In addition, the Innu Nation received First Nations Status in 2002, the last First Nations group in Canada to do so. The gaining of status has opened up negotiations for self-government (right of use to Innu), but has also increased the need for educated and skilled Innu negotiators, lawyers and scientists, and sound ecological information.

The Innu have considerable self-interest in facilitating scientific research in Labrador. They are in the process of negotiating a land claims agreement with the federal government for which they have to do extensive documentation of their own land use. As they win a voice in environmental assessment and impact–benefit agreements for specific development projects, they need rigorous baseline scientific data. From the perspective of the Innu Nation, part of the challenge has been finding scientists and researchers willing to work with Innu priorities, not the other way around. In the words of Peter Penashue, past President of the Innu Nation, it has been a 'long hard road' toward achieving self-determination since the formation of the Innu Nation twenty-five years ago:

> From the start, we worked closely with anthropologists and other scientists to help us build our case. As many of you know, Canada takes the position that Aboriginal people who wish to negotiate land claims must prove that we were here first, before the Europeans.
>
> Now, what seems like common sense to us is obviously hard for governments to understand, so we had to hire experts to work with our *tshishennuat* to conduct interviews in order to make the maps that were finally accepted by governments in 1989. Throughout this process, we had to explain to our governments why Innu were the ones who had to make the maps, not the governments … That kind of relationship also exists between Innu and many scientists.
>
> We've anticipated many environmental assessments, and we've often been puzzled by the certainty of some of the experts that governments and companies bring up from the south to tell us about our land and the animals that we have studied for thousands of years. It was frustrating to listen to some consultant who maybe had spent a summer in our territory, or more likely had read a few reports about it, think he understood our land better than our *tshishennuat* who had spent their entire lives there. It shouldn't be surprising, then, to hear that many of our people started to think that some scientists would say anything if they were paid well enough. These kinds of scientists would descend on our communities every summer and leave them again in the fall. They were kind of like flies. They buzzed around, getting into people's hair, but we thought that they were mostly just an annoyance. This is until we realized that governments took these scientists seriously, and used their findings to approve developments like low-level flying.
>
> What scientists often forget is that Innu, like everyone else, have priorities …. We spend so much of our time trying to come to terms with the White man's world that many of our people have lost touch with their own. Regaining control of our lives, our communities, and restoring our culture are among the most important goals. (Penashue 2001: 4–6)

From the Innu view of the world, Western science is based on a mechanistic conception of the universe. It views the world as composed of discrete entities and processes, and is biased towards studying things that can be quantified and measured (Sillitoe, this volume). The Innu worldview, as with many aboriginal people, features a more holistic approach to the world, inclusive of the people who inhabit it. They see themselves as an integral part of the landscape, and are wary of their knowledge being used in a way that betrays this fundamental principle. 'Science' is not seen as a neutral framework, but rather as highly political. Considerable power lies in the hands of those who decide what information will be gathered and how it will be used (Innu Nation 1999: 2–3).

Over time, the Innu Nation has developed its own agenda and determined its goals in accordance with their needs. What has made the *Ashkui* Project unique has been that it is Innu driven and conducted in dialogue with the Innu Nation. As noted by Peter Penashue:

> Today, most of the researchers who come to our communities are invited by the community itself. They generally work as part of the agenda the Innu have helped to develop. This has allowed us to deal with developers on our own terms, and challenge their science with our own science …The *Ashkui* Project was developed out of the dialogue about these issues that has taken place for several years between researchers and members of our community. This is a reversal of the usual research relationship between Aboriginal communities and researchers, where the researchers would come in from the outside with a project already conceived, and attempt to sell it to the Innu. (Penashue 2001: 5–6)

While the Labrador Project has been sensitive to Innu priorities, e.g., the *Ashkui* Project was based on the Innu Nation's suggestion of *ashkui* sites as a research focus for Environment Canada, it has still been a challenge for both sides. To access IEK, western scientists have to appreciate a whole different way of conceiving of the world. It requires scientists to open up their scientific paradigm to accommodate Innu knowledge, and vice versa. From the Innu side, it is difficult to translate scientific concepts into their language when no corollaries exist. For instance, when various scientists from the federal government explained their mercury research at a community meeting in the Innu community of Sheshatshiu, Labrador, it was pointed out by an Innu Guardian working with Environment Canada that the Innu had no concept of mercury or its relevance to their community. The scientists had assumed they were speaking in universal language and that mercury was a commonly shared concept.

At all stages, the entire project depends on translating from one knowledge paradigm to another without losing the integrity of either. An inevitable knowledge hybridisation will occur, but this can be a positive outcome if it is a chosen and conscious process rather than a choiceless 'mental colonization' (Battiste 1998: 20) and the process involves all parties (Dove et al., this volume). To set up this dialogue has required a shift in thinking and in practice on the part of Environment Canada staff as well as the Gorsebrook Research Institute social scientists. The shift is not only in who drives the agenda, but also in who controls the knowledge, and how that knowledge is collected, archived, published and made meaningful to the Innu.

Environment Canada

Environment Canada, Atlantic Branch has been interested in Labrador since the early 1990s. The area represented a gap in the agency's knowledge of ecosystems in Atlantic Canada, and accounts for over 53 percent of the region's territory (approximately 286,000 square kilometres). The ecosystem in Labrador is threatened by a number of anthropogenic activities, including clear-cutting, nickel-mining, low-level flights from a NATO air force base, hydroelectric projects, commercial fishing and climate change. Environment Canada wanted to gather baseline ecological data in order to monitor the environmental impact of these activities and to ensure compliance with environmental agreements and legislation on the local, national and international level. As well, the relatively pristine conditions provided a natural laboratory to monitor environmental change.

Initially, Environment Canada had a narrow interest in wetlands classification. It attempted to gather information using satellites, airplanes and a crew of scientists on the ground, without consulting the local community (Wilson 2004). When this approach proved unsuccessful in obtaining community buy-in, Environment Canada sought input from the community. At the suggestion of an Innu contact, four staff members went on a cultural orientation trip 'in-country' during the winter of 1997, tenting for five days in −25°C weather. This camping experience fostered a profound appreciation of the Innu culture and their relationship with the land. From these early consultations with members of the Innu community, *ashkui* sites were identified as culturally significant to the Innu way of life, particularly in the springtime when families camped in tents near *ashkui* for weeks at a time. Research at these places would benefit both parties. As mentioned previously, the sites allowed for Environment Canada's continued focus on water quality and waterfowl inventories and seabird research. And it could answer Innu concerns about changes in the environment from development projects, e.g., NATO low-flying jet corridors over *ashkui* sites; acid rain from the south; mercury levels in the water, mammals and fish from industrial developments; and climate change, to mention a few.

Following these initial consultations, Environment Canada invited social scientists at the Gorsebrook Research Institute to join the project to help 'develop the capacity to recognize the intrinsic values of Northern landscape and peoples' (Environment Canada 1999: 3–1). The agency needed assistance in translating between IEK and Western scientific knowledge, an area with which researchers at the GRI had experience. In May 1998 the first GRI social scientist went up to Sheshatshiu, Labrador and spent over three weeks in the community with *tshishennuat* and Innu co-researchers/Guardians and translators documenting environmental knowledge. Eight days were spent at an *ashkui* site at Lake Shipiskan with an Innu family who had camped there for generations.

The research conducted during this initial and subsequent field trips not only served to document extensive land use practices of the Innu (subsequently included in Environment Canada's on-line maps) but also identified a number of *ashkui* sites that were in current use or had been used in the past by Innu families. Environment Canada scientists proceeded to set up the *ashkui* research network stations in the spring of 1999 and hired an Innu co-worker (and future Environmental Guardian) to assist in the research, testing and presentation of results to the community. This network was designed to provide information on *ashkui* formation and behaviour across a latitudinal gradient. The sites were sampled for eighteen parameters (major ions, metals and nutrients) from the first opening in the spring until the summer or early fall. Along with conducting water quality sampling to establish seasonal patterns, questions of particular importance to the Innu people were addressed, such as site productivity, sensitivity to acid rain and drinking water quality.

Environment Canada found this nascent partnership beneficial. It noted in its 1999 evaluation report that the Innu were able to add new insights into the description of the environment, and that they could decrease the cost of fieldwork by directing and focusing the research (Environment Canada 1999: 5–6). This laid the groundwork for the inclusion of local human resource capacity building in the project mandate, an evolution that met with some initial resistance within Environment Canada because of the shift from its normally mandated scientific responsibilities towards the inclusion of social science work (Wilson 2004). Including the Innu also had a significant effect on the project's philosophy and direction. For instance, the 1999 Environment Canada report discusses the shift towards a more holistic, ecosystem-based approach and acceptance of the principle of 'biological integrity' as a result of Innu participation (Environment Canada 1999: 4–5, 6). Unexpected spin-offs, such as NATO low-level jets altering their flight corridors to avoid the waterfowl staging areas, resulted from the *tshishennuat* identifying productive *ashkui* sites, which, in turn, led to Environment Canada designating these sites as waterfowl staging areas.

Why are *Ashkui* Important?

When asked what *ashkui* means, most Innu say it refers to open water in the ice (literally, clear water). But rephrasing the question to ask, 'What do people do when they camp at *ashkui* in the springtime?' opened up a description of a way of life that involved a web of relationships between people, animals and the landscape. These descriptions involved the worldview and personal histories of Innu families. These histories are embedded in the landscape where Innu have camped and lived for generations. Sites associated with legends throughout the landscape further attest to the inseparability between the Innu psyche and the land. It is for this reason that Innu speak of the land as 'medicine', and liken *ashkui* to supermarkets and pharmacies, where you can get everything you need.

This intimate and personal relationship with the land is borne out further in the *tshishennuat* description of *ashkui* as 'like a father because it provides everything'. Many *tshishennuat* also talked of ice as a living being. It has its own sounds, and a person has to learn how to listen to, talk to, and make offerings to ice in order to survive and travel safely. The legends of 'ice person' inevitably ended in misfortune if people abused ice. Some people had been told stories as children that the ice would grab them if they got too close to it. Drownings were always a concern near *ashkui* sites, especially for children playing too near the thinning edges of the ice in the springtime while their families camped near *ashkui* sites.

Because of their longstanding relationship to their ancestral landscape, the *tshishennuat* are sensitive to the many environmental factors that affect the break-up of ice and the opening of water. These conditions vary from one body of water to the next, from one day to the next. *Ashkui* can occur on any body of water – lakes, rivers, ponds, and even puddles. Winds, rain, currents, depth of water, the presence or absence of rivers and brooks flowing in and out of lakes, the weather, as well as man-made alterations to the environment, are all factors that affect the formation of *ashkui*. *Ashkui* are present in the fall as ice forms, and in the spring when ice begins to break up. In some places, perennial freezing over never occurs because of the continual shifting of ice caused by the fast currents.

During the initial research, three types of ice were mentioned in discussing the formation of *ashkui*. The first was white ice, or solid frozen ice. The second was black ice, also called rotten ice, which appears as the ice begins to melt and push up from below the surface. The third was the 'nail' ice or a crystal-type ice that has the appearance of nails or crystals as the ice begins to melt. The ice below darkens as it begins to melt, and then pushes upward, or pops up. This occurs after the edges of the lake open and *ashkui* have begun to form. At this point, the winds play a crucial role in moving it off the lakes. It is the winds that move the ice back and forth on the lakes causing it to break up and move off down rivers if present. Another impact of the movement of ice by the winds is the piling up of the ice on the lake shores.

The ability to predict weather is obviously important. Winds, rain and sun all affect the ice conditions and the ability to travel on land and water, as well as the presence of animals around *ashkui*. Three birds were mentioned as weather indicators: the loon, the robin and a bird the Innu refer to as the rain bird. When the loon sings, it forecasts winds and storms. The robin foretells rain. The rain bird, which has the appearance of a swallow and arrives in the spring, forecasts rain when it sings. Others spoke about using the stars, the sky colour, and the way smoke rises as methods of predicting the weather. Other indicators are shooting stars, which tell which way the wind will blow from the direction they fall, and the redness of the sky at night and in the morning. Although most people do now have two-way radios in their camps to keep posted on news and weather, reading the weather from environmental indicators remains significant.

Ashkui sites are associated with the first arrival of flocks of waterfowl in the spring drawn to the open water. Later in the spring, the birds will pair off for mating and the laying of eggs in the nearby marshes, woods and islands. These wetlands, islands and wooded areas are also a significant part of the overall *ashkui* ecosystem. It is near these *ashkui* sites that Innu families set up camp and live for weeks and months at a time to exploit the rich and varied resources of the land and water. The sites are also significant for the fish that come to the newly open water and the numerous animals that come near *ashkui* to eat fish or feed off the plant material washed down by the fast spring flows of the rivers. The smaller openings at the mouths of brooks are where animals come, and are favoured places for trapping a number of small mammals; they are also favoured caribou crossing points. During interviews, people were shown the pictures of waterfowl and were asked to discuss when they arrived, special characteristics, eating habits (vegetation vs. fish, deep divers vs. dunkers or swimmers), their habitats, and the uses to which different parts of the birds were put. Each picture was labelled with the Innu name for the waterfowl, and informants were asked the meaning of these names, and to make corrections.[4]

Inevitably, climatic and environmental changes have affected the formation of *ashkui*. Many *tshishennuat* mentioned that *ashkui* were opening earlier, or that springs are shorter at their camping areas than in previous years. At these places, *ashkui* would usually form in the first week of May, although one estimate was that *ashkui* were opening one month earlier than twenty years ago at one lake. Ducks and geese are also arriving sooner with the earlier opening of *ashkui*. One *tshishennua*[5] estimated that thirty-five years ago there was still ice in May and even June, and people would not go to camp near *ashkui* until early June. Now there is no ice by the end of May. Further, increased in rainfall creates more weight on the ice as it collects on the surface, which presses down on the ice thus expediting the popping up of the ice from below the surface.

Others attributed earlier spring break-ups to the Churchill Dam. One couple who camped in the Churchill Falls area mentioned that the ice sits higher on the water, no longer touching it, and that it is really dangerous for people who do not know about *ashkui*. There are now more *ashkui* everywhere. Sightings of Harlequin ducks had also diminished, according to one woman, and had not been seen since the building of the dam. Another couple complained that environmental changes are reflected in less fat on the animals because they work hard looking for food to eat, and the food is not quite right in the winter. A once productive *ashkui* site at Northwest River was destroyed by the building of a bridge. This site used to be rich in seals, ducks, and fish, including salmon. Seals came up onto the ice edge as the water opened. Seals were once of major importance to the Innu, used for medicine, food and clothing. They would cook the fat and drain it into a bottle or can and take it as medicine. Sealskin boots were especially good in *ashkui* when there was slush because the water did not penetrate the hide.

Low-level flying from the NATO air force base in nearby Goose Bay was mentioned during a number of interviews. People had first-hand experience of low-flying jets and described the effect on animals and fish, and how frightening the sound was to people and animals. Stories of children and women running scared during an overhead pass were told by a number of people. Dead fish also started appearing in the water. And beaver were hesitant to come out for their night-time feeding during night-time flights. Many *tshishennuat* also spoke of how the taste of the meat from animals had changed particularly when hunted near industrial development sites, and mentioned observations of enlarged hearts, or other malformations in animal and fish.

The collaborative research effort has amassed a large body of information, now digitised and incorporated into maps and environmental assessments. As an unexpected spin off, this research assisted the Innu in their much publicised battle to change low-level NATO air force flight-testing corridors passing over *ashkui* sites, upsetting the life below. Currently the construction of an extension of the Trans-Labrador highway through Innu lands is further focusing interest on a specific area of wetlands and migratory bird habitats, staging grounds and moulting sites in a unique baseline study of environmental impact of the highway.

Western Scientific Perspective – Making it Relevant and Addressing Concerns

The objective of the Western science research was to undertake water quality investigation. The first objective was to develop a basic understanding of the water quality and general limnology of Labrador freshwater ecosystems. The second objective was to evaluate the trophic status of *ashkui* and to relate productive potential of these sites to the Innu observations of them as areas which are important for fish and waterfowl production. The *ashkui* sites are sampled yearly starting from initial open-water formation, through the summer period and into the autumn. At each *ashkui*, field measurements of water temperature, specific conductance, dissolved oxygen, turbidity and pH are taken. The water is also tested for ultra-trace mercury.

The focus of the *Ashkui* Project on the value and importance of different cultures, activities and perspectives on the Labrador landscape requires a dedicated emphasis on reporting results in ways that have meaning across cultures. In an effort to make the water chemistry research at *ashkui* meaningful to the Innu people, Environment Canada scientists endeavoured to answer some of the basic questions that the *tshishennuat* and other Innu asked them. The following are some of the questions that Environment Canada Scientists attempted to answer about water quality at *ashkui* sites. These questions came from the immediate concerns voiced by the Innu regarding the changes they were experiencing in their environment and issues fundamental to their way of life.

Will it Make Good Tea?

Tshishennuat often talk about *ashkui* water from the perspective of the taste of tea. The ability of the *ashkui* water to make good-tasting tea could be evaluated using a variety of water quality measures. Most of the *ashkui* had good quality water although some sites with high water colour (e.g. Susan River) or marine influence (e.g. Carter Basin) would be less suitable for making good tea; for example tshishennuat camping at the Susan River *ashkui* talked about poor tasting tea there. (Figure 6.1 indicates answers to questions.)

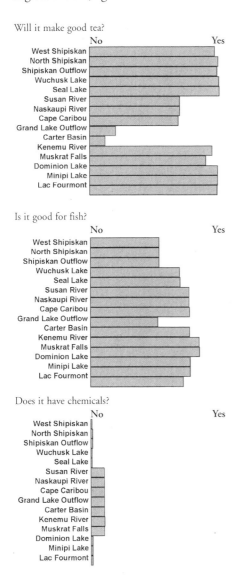

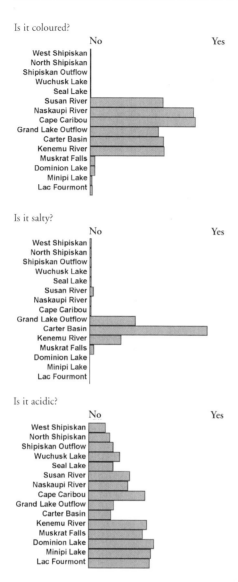

Figure 6.1 Assessment of *ashkui* water quality.

Is it Good for Fish?

The ability of an *ashkui* to support healthy populations of fish and other animals can be predicted from simple measures of primary production. In general, all of the *ashkui* have low production which suggests that they will support small but healthy communities of aquatic plants and animals. *Tshishennuat* have remarked that the concentration of fish at *ashkui* is thought to be related to high water movement and available light at open areas. The low production estimates also suggest that physical aspects may be responsible for high fish abundance at *ashkui*.

Does it Have Chemicals?

One of the major concerns expressed by the Innu is whether or not the water is safe to drink. The metal concentrations at *ashkui* are low, suggesting that in general the water is safe for human consumption. In the more highly coloured waters, concentrations of metals such as aluminum are higher but this is to be expected and is not a health concern. The chemical analysis does not include evaluation of bacterial contamination, although given the remote location of these sites, elevated bacterial populations are not expected.

Is it Coloured?

Waters which drain bogs and other wetlands are often coloured brown by decaying plant materials. As decaying plant material produces fulvic and humic acids, these tea-coloured waters often provide less favourable habitat for fish and other aquatic animals and are less appealing as a source for drinking water. The *ashkui* in the Grand Lake area are the most highly coloured while the northern and southern sites are generally clear.

Is it Salty?

Tshishennuat appear to have considerable capacity to identify changes in salt content of fresh waters. These changes in salt have a profound influence on perceptions of the use of water for drinking. For example, several *tshishennuat* have talked about Grand Lake water near the Susan River being saltier now than in the past. Interestingly enough, sodium and chloride are often the dominant ions at the Susan River site. Several sites (Grand Lake Outflow, Carter Basin and Kenemu River) have salt influence from Lake Melville.

Is it Acidic?

Tshishennuat have expressed concerns about acid rain. The dilute nature of waters in Labrador makes most sites sensitive to acid rain, as they have limited ability to counteract incoming acids. However, none of the sites have as yet experienced major acid rain impacts. Excluding the sites influenced directly by salt water from Lake Melville, the potential for acid rain effects is more pronounced in the southern sites. This makes sense as these sites are located closer to the industrial sources of acid emissions.

In essence, bringing the research back to the community is a way of reaffirming the Innu's practical knowledge from a scientific perspective, and showing the importance of their roles as collaborators in the project design and assuring that their knowledge features in and is passed on to policy makers. It gives Innu Knowledge political credence and significance which is central to them achieving a degree of autonomy in their own country. In turn, this will aid Innu in addressing increasing resource management responsibilities emanating from land claims negotiations and related co-management and benefit-sharing agreements, such as ones recently signed with the province of Newfoundland and Labrador Department of Forestry Resources and Agrifoods and with the Voisey's Bay nickel-mining operation taking place within Innu ancestral lands.

Labrador Project Evolution: Innu Environmental Guardians Program

A rich and comprehensive body of knowledge has been compiled that is of mutual benefit to the Innu community and environmental scientists. The work has evolved into assisting the Innu with a community capacity-building project referred to as the Innu Environmental Guardians Program (IEGP). The aim of the programme is to develop an educational path to train Innu in the management and protection of their ancestral lands based on Innu traditional values and current community needs. The program requires education not just as it is defined by Western scientists, but as it is defined by *tshishennuat*, custodians of Innu environmental knowledge and worldview. The Environmental Guardians concept recognises the importance of both the longstanding and substantial body of environmental knowledge held by the Innu, and the need for the Guardians to develop competency within Western scientific and technical disciplines concerned with environmental protection, management and resource use. Incorporating these two ways of knowing requires Innu Environmental Guardians to acquire a unique set of skills and competencies that can reflect both Innu knowledge traditions, and the disciplines and skills that are recognised by formal Western educational institutions. At its core, however, the programme is based on Innu values, needs and concepts of well-being.

The key to the programme's success is incorporating on-going community concerns, with which the Guardians are involved daily. Some of the programme's key components are:

- Courses offered in 2–3 week modules and delivered within the community or at field sites where projects are underway.
- Learning is related to on-going projects, e.g., a forestry co-management agreement with the provincial government, the monitoring of waterfowl and wetlands in relation to the building of the Trans-Labrador Highway

extension, monitoring the Voisey's Bay nickel-mining activities, etc.

- Modules are scheduled around 'real' life situations, e.g., seasonal work, family obligations, and time at camps in the country.
- Training crosses disciplines, Guardians receiving training in all aspects of environmental monitoring.
- Programmes are bilingual (Innu Aimun and English) when *tshishennuat* are present.
- The programme is Innu-driven, which means the community decides the priorities, and learning is geared towards the preservation of their own land use and cultural practices. This knowledge is then used as the basis of decision-making processes in any development project.
- *Tshishennuat* are involved as advisers and as teachers.

The Guardians play a key role as translators and communicators to and from their communities, and within their own communities between different generations. Many government agencies, educational institutions, businesses and nonprofit organisations are approaching the Innu for a variety of research and development projects. The Guardians are responsible for communicating their cultural beliefs and values to these various outsiders, and then relate what is discussed back to their community in a meaningful and comprehensible way. This translation process back to the community involves the creation of new terminology for scientific concepts that have no equivalent in Innu Aimun (Innu language) since many of the *tshishennuat* speak only Innu Aimun. In short, the Guardians provide access to information often inaccessible to community members, particularly *tshishennuat*, and also create intergenerational bridges. This communication helps unify the community as well as address the loss of knowledge that will occur with the passing of the *tshishennuat*.

As a team, they support one another, share knowledge, offer different strengths to solve problems, and so forth. In turn, the community is beginning to regard them as a team of experts with a mandate to manage and protect their lands and cultural heritage. This includes the *tshishennuat*, who through their involvement are respected as teachers, but who are also learning the new way of looking at the environment through Western scientific eyes, and being able to compare it to their own ways. Increasingly, a hybridisation of knowledge occurs and the question that can only be answered in time is how this knowledge will be carried forward to serve the well-being of the community in the generations to come, given changing political winds and new generations yet to born.

During the autumn of 2002, some members of the Saint Mary's University faculty joined in the effort to develop a curriculum, offering formal academic accreditation to the training modules as a way to open up a university path to the Guardians. There was no requirement for any Guardian to take the two modules offered for credit – the intention was to train the Guardians, whether or not credit was gained. However, during the first of a two-module accredited course, all fourteen participants signed up with the University. This is a unique and historic move. Saint Mary's University faculty acknowledged the role of the *tshishennuat* as the legitimate authority on their own knowledge and teachers of equal stature on the modules. During the first accredited course, the *tshishennuat* were involved in the evaluation of the Guardians.

The *tshishennuat* are co-evaluators with the module instructors. The Guardians are required to present their learning to the *tshishennuat*, who then evaluate according to their criteria. In so doing, the *tshishennuat* are learning about such concepts as ecosystems, watersheds, urbanisation and pollution, as well as the use of instruments such as the GPS, clinometers, and compasses. Anecdotally, during an evaluation where the Guardians were displaying the use of various instruments they use on the land, a *tshishennua* said, 'We carry GPS in our heads to find locations.' This was a poignant remark because it showed both his understanding of how the GPS worked as well as the changes occurring in how environmental knowledge is studied and stored.

Some of the faculty at Saint Mary's University also recognise the need for people to write and present in their first language within the Western educational system. Faculty working with the Innu Environmental Guardians Program are attempting to broaden the definition of literacy to recognise different languages as unique and enriching to the educational process. Doing so does not overlook the necessity to learn English as a requirement for dealing with contemporary issues, but English does not have to be exclusive of other languages. Recognising the importance of including the *tshishennuat* as well as responding to community needs, especially by traditionally established educational institutions, is part of the process to develop trust across cultures.

Although these are steps towards redressing the legacy of distrust brought by settlement and colonisation, there is still much more ground to till. Some of the Guardians, succumbing to social pressures and the recent influx of money into the commuity from Voisey's Bay, have been laid off, while Environment Canada itself has been undergoing reorganisation, shifts in priorities and funding cuts that are impacting on the programme. With the death of Geoff Howell, a crucial player in and protector of the project from the beginning, the value of the project began to be questioned by people within Environment Canada regarding the role of capacity building within scientific research.

Conclusion

Capacity building has to involve all stakeholders. Putting Western scientific and Innu knowledge in dialogue provides a much more comprehensive understanding of the Labrador landscape, and has raised Environment Canada's awareness of the significance of culturally valued areas. Through enhancing the role of the Guardians, they, in turn, can provide the community with access to information, allowing for greater community in-put for decisions affecting their ancestral lands. Working with *tshishennuat* as advisers and professors bridges intergenerational divides that have arisen since settlement, and assists in the incorporation of Innu environmental knowledge into resource management, stemming the loss of knowledge with the passing of the *tshishennuat*. Further, the incorporation of Innu knowledge and the use of Innu Aimun within the educational process expands the notion of literacy, including scientific literacy, commonly held within university settings. This recognition of Innu Aimun adds further credence to language preservation efforts as an integral part of resource management and environmental sustainability. The *tshishennuat* involvement with the learning process of the Guardians makes them privy to the information exchange, and empowers them as professors of their knowledge equal to that of university professors. The offering of the choice for academic accreditation by the university opens a pathway to higher education that many Innu have rejected, or have been rejected from, entering in the past. Further empowerment of the Guardians came through the dedication of an Environment Office from the Innu Nation, giving the Guardians a distinct space from which to operate, again fostering an identity as a team of people versus individuals hired independently by various government agencies, as had been the case in the past.

In a preliminary way, we suggest six indicators for wholeness or wellness of the people in relation to any project. Each of these questions could be developed into an indicator, or a measurable value.

1. Have all the people been engaged in defining the motivation to undertake the project?
2. Does the research serve the community as well as the investors? Who is the ultimate beneficiary of change?
3. Who is defining the knowledge that is being gathered and documented? Is it inclusive of all stakeholders?
4. Who is governing the decision-making process and to what ends?
5. To what extent have avenues of communication, e.g., different languages, been included and respected?
6. To what extent have cultural land use practices and values been included in co-management agreements?

The true significance of the research is in the relationship between the collaborators, and the way research is conducted so that Innu priorities are respected and their needs for building capacity met. It requires a holistic view of the landscape, an integrated, interdisciplinary approach to knowledge,[6] and a deep appreciation of the 'translation' process and role of the 'translators'. It has shown that communities are willing to take on these roles, develop a hybridised language to work with issues, and step forward into decision-making roles to bring about policy changes.

Notes

1. Geoff Howell died unexpectedly in 2003, leaving the Askhui Project as part of his legacy. It was his initial vision to work collaboratively with the Innu Nation and social scientists at the Gorsebrook Research Institute to create a new paradigm for scientific research. Since his death, many changes have occurred in the project.
2. Dave Wilson, former project manager for the *Ashkui* Project at Environment Canada, Atlantic Region, retired in the spring of 2004. The Innu Nation, in gratitude for his years of dedication to the project, presented him with a gift. The gift was a painting of one of their respected *tshishennua*, Matthew Penashue, now deceased, accompanied by the story of Matthew's life. The picture was painted by Mary Ann Penashue, wife of the past Innu Nation president, Peter Penashue.
3. Anthropologist Frank Speck in writing about the Innu (formerly referred to as the Montagnais-Naskapi) stated, 'No wonder, then, the proper food of the tribe being either directly wild fruits or indirectly vegetable through the diet of game animals, that with their food in whatever form consumed, the Montagnais-Naskapi are "taking medicine." Thus the native game diet is prophylactic to mankind. A deep significance lies beneath this doctrine' (Speck 1935: 81).
4. Much has been written about the methodological weakness of using pictures as a research tool for developing indigenous taxonomies (Sillitoe 2003). In this case, a respectable body of research had already been developed regarding Innu taxonomies, e.g., Clement (1995) and the ongoing work of linguists, José Maillhot and Dr Marguerite MacKenzie, and my research interest was not in Innu taxonomic systems. The emphasis in the interviews was on the habits, conditions and characteristics of the different species of waterfowl that came to *ashkui*, the relationship of these species to *ashkui* in terms of the Innu way of life and worldview, and changes over time that could be identified in their behaviour.
5. *tshishennua* is the singular form of *tshishennuat*, according to José Maillhot, linguist and recognised authority on Innu Aimun (Innu language).
6. I recognise that there is a body of writing relating to interdisciplinary research (Sillitoe 2004) and the challenges involved. This is a topic that needs further consideration in the work we are undertaking, particularly with some shifting political and funding structures that are currently affecting our project.

References

Battiste, M. 1998. 'Enabling the Autumn Seed: Toward a Decolonized Approach to Aboriginal Knowledge, Language and Education', *Canadian Journal of Native Education: Creating Power in the Land of the Eagle* 22(1): 16–27.

Bieder, R.E. 1986. *Science Encounters the Indian, 1820–1880: The Early Years of American Ethnology*. Norman, OK: University of Oklahoma.

Canadian Biodiversity Strategy. http://www.eman-rese.ca/eman/reports/publications/rt_biostrat/cbs27.htm.

Canadian Environmental Protection Act (1999), available at http://laws.justice.gc.ca/en/C-15.31/text.html.

Clément, D. 1995. 'Why is Taxonomy Utilitarian?', *Journal of Ethnobiology* 15(1): 1–44.

Convention on Biological Diversity, Principle 8.j. http://www.biodiv.org/convention/articles.asp?lg=0&a=cbd–08

Davis, W. 2002. 'Wade Davis: Writer and Defender of Life's Diversity', *Ideas CBC Radio*, 4 February.

Environment Canada 1999. *Labrador Biodiversity and Environmental Monitoring Project: Linking Western Science and Indigenous Ecological Knowledge*. Halifax, Nova Scotia.

Innu Nation 1999. *Innu Nation Discussion Paper: Research Priorities in Nitassinan, Four Perspectives on Ecosystem Research Priorities in Labrador*, (ed.) Environment Canada, Collaborative Research Workshop, Memorial University, St. John's, Newfoundland, 8 April.

Penashue, P. 2001. 'Building Respectful Research Partnerships', in *Ashkui Symposium Papers*, compiled by Trudy Sable. Halifax, Nova Scotia: Gorsebrook Research Institute.

Rio Declaration on Environment and Development, Principle 22 available from http://www.unep.org/Documents/Default.asp?DocumentID=78&ArticleID=1163.

Sillitoe, P. 2003. *Managing Animals in New Guinea: Preying the Game in the Highlands*. London: Routledge.

——— 2004. 'Interdisciplinary Experiences: Working with Indigenous Knowledge in Development', *Interdisciplinary Science Reviews* 29(1): 6–23.

Speck, F.G. 1935. *Naskapi: the Savage Hunters of the Labrador Peninsula*. Norman: University of Oklahoma Press.

Wilson, D. 2004. Personal Interview, 26 March 2004. Environment Canada Atlantic Region Office, Dartmouth, Nova Scotia.

7 Globalisation and the Construction of Western and Non-Western Knowledge

Michael R. Dove, Daniel S. Smith,
Marina T. Campos, Andrew S. Mathews, Anne
Rademacher, Steve Rhee and
Laura M. Yoder

Introduction

Anthropological interest in non-Western knowledge dates from the very beginning of the discipline. Early anthropologists interested in the so-called 'savage' or 'primitive' mind asked, in effect: do non-Western peoples think differently from Western peoples and, if so, how? In the years since, this question has periodically surfaced, been critiqued, submerged and reappeared. A recent incarnation – and one of particular importance to the reigning paradigms of global conservation and development – involves non-Western, indigenous environmental knowledge.

Anthropological interest in indigenous systems of resource management also dates back to the early years of the discipline, and especially flourished with the rise of ethnoecology in the 1950s. In the 1960s and 1970s, anthropologists began to invoke indigenous systems of knowledge and practice to critique the dominant development paradigm and its privileging of extra-local knowledge. As the sustainability of many resource-use systems built on the Western scientific paradigm became increasingly suspect, the pervasive deprecation of non-Western resource management was replaced, even among some practitioners in conservation and development, by valorisation.[1] Whereas this about-face represented a useful correction to earlier views, the underlying division between Western and non-Western systems has come increasingly to be seen as problematic theoretically (Agrawal 1995; Dove 2000).

This East–West division is characterised by two major paradoxes. First, the recent reification of indigenous, non-Western knowledge has occurred just as anthropological theory developed over the past generation has gone in the opposite direction. Much of this theory has focused on deconstructing conceptual boundaries that had hitherto been central to the discipline, including the boundary between indigenous and non-indigenous. For example, one of the principal accomplishments of the 'world system' studies of the 1980s was to demonstrate how all of the world's peoples – even seemingly isolated societies – have been caught up in global history (Wolf 1986; Wilmsen 1989). Over the past decade, another line of study has focused on critiquing traditional anthropological assumptions of the integrity and homogeneity of local communities, emphasising instead the potential for internal differentiation and conflict (Li 1996; Mosse 1997).[2] Both developments helped to promote a theoretical interest in hybridity, which is almost the antithesis of indigeneity, as one of the defining characteristics of the modern, postcolonial era (Gupta 1998).

For these reasons, among others, some anthropologists are critical of the current deployment of the concept of 'indigenous people'. Kuper (2003) published a critique of the empirical basis for claims to indigeneity, which set off a lengthy debate of the possible negative implications of this critique for disadvantaged groups trying to trade on their indigeneity. Less publicised and contentious but equally challenging are the arguments that claims to indigeneity are themselves politically fraught because they are too narrow or exclusive. Conklin and Graham (1995), Li (2000) and Stearman (1994) argue that the concept of indigenous environmentalism is so exacting and constraining as to be disadvantageous in the long run to groups that adopt this label. Gupta (1998) argues that the concept of indigeneity, while potentially self-empowering for some groups, is inaccessible to the vast proportion of the less-developed world.

The prospects for self-empowerment, however narrowly defined, lead to a further debate about artifice. Stearman (1994) has argued that indigenous, non-Western environmentalism must be self-conscious if it is to be reckoned as true conservation; while Ellen (1999) has looked at how self-conscious conservation practices might emerge. But since this same self-consciousness is interpreted by some observers as evidence of inauthenticity and perhaps even opportunism, Li (2000) sought instead to represent indigenous claims as 'strategic essentialism' (following Spivak) or 'articulation', a process not of opportunism but of simplification and boundary making.[3]

These possibilities of genuine and non-genuine indigeneity bring us to the second paradox: modernity both constructs and erases the Western/non-Western divide. This paradox turns largely on the dual nature of construction and erasure as simultaneously symbolic and material. Modernity symbolically constructs indigeneity, otherness and the idea of fundamental difference between Western and non-Western societies even as it promotes the social and material dismantling of whatever divide may once have existed through the mixing and hybridisation of elements that lie on either side of it. Scott (1998) has argued that modern or high-modern thought is not merely a reaction against locally specific traditions and ways of thinking, but is also dependent on these traditions as a foil against which to define itself.[4] But the local and the traditional against which the global and the modern defines itself is also a product of modernity. Some scholars now argue that the reification of locality and identity that is associated with 'indigenous' or non-Western society is itself a product of modern histories of engagement and confrontation. Benjamin (2002) writes that all reported tribal societies are actually 'secondary formations', which developed as a result of steps taken to stay apart from the state (cf. Li 2000).[5] Moreover, Mosse (1997) suggests that 'production of locality' is one of the explicit goals of modern development policy in many countries, because locally autonomous, internally sustained and self-reliant community institutions not only meet current ideals of community-based management but also meet state imperatives of cost sharing, recovery and reduced financial liability.

Hirtz (2003) has taken this argument a step further to argue that indigenous communities are only able to articulate their identity through the use of modern administrative procedures and representational processes. It is through the very process of successfully articulating an indigenous identity that these groups most definitively enter the realm of modernity, which leads to Hirtz's paradoxical conclusion that the ability to assert indigeneity means that one is not indigenous.[6] Evidence of the imbrication of indigeneity and modernity notwithstanding, indigenous values and identity have been widely valorised in recent years precisely because they appear to be not only non-modern but anti-modern. Giddens (1984) and Appadurai (1996), like Scott (1998), have argued that the modern global system is 'disembedding', that it destroys locality. This is deemed by many observers to have ill consequences for society and environment alike, the solution to which is thought to be restoration of locality or re-embedding (Hornborg 1996). This explains the rise of interest in indigeneity, which is thus both a product of, and a marker of, modernity.

The relationship of indigenous, non-Western peoples and knowledge systems to the modern world system is thus ringed with paradox (Sillitoe 2002). One way to gain insight into the workings of this paradox is through ethnographic case studies of the global 'circulation' of specific bodies of knowledge. Recent contributions to the literature on the social construction of global environmental problems focus on how environmental knowledge is transported and transformed, in particular between Western and non-Western societies (e.g., Rangan 1992; Brosius 1997; Gupta 1998; cf. Zerner 1993). Environmental concepts do not 'travel' independently from one place to another and impose themselves on agency-less people. Rather, the concepts of one part of the global community are appropriated, transformed and contested by specific local actors when they move elsewhere, for which reason the term 'deployment' or perhaps 'mobilisation' of ideas might be preferred to circulation (Tsing 2000, 2005). Concepts become powerful in a new setting only if they can be integrated into it, at the same time as a part of their power derives from continued identification with their place of origin. The elasticity of the transported concept is thus a key to its collaborative use by different parties. Indeed, Tsing (2000) maintains that successful global coalitions depend upon the mis-translation of ideas.

These issues will be further explored here in an analysis of three dimensions of the global mobilisation of environmental knowledge, namely the construction of alterity, the simultaneous, yet often obscured, process of hybridisation, and how both of these processes come into play in the strategic use of indigenous identity. We have already discussed the thesis that indigeneity is constructed alterity, and that such constructions have both a potential up-side and potential down-side for the peoples involved. We will illustrate this point with two case studies of how 'otherness' is constructed, one involving Third World swiddens and the other involving the so-called 'Northern Forest' of the Northeastern United States. The second and opposite (yet closely related) dimension of global environmental discourses to be explored here is their hybridity, the mixture not only of Western and non-Western knowledge traditions but also of formal scientific as well as folk epistemologies. Here we will examine case studies of the mixing of scientific forestry and peasant knowledge in Mexico's Sierra Juarez mountains and the adoption and adaptation of Western cartographic traditions by indigenous Third World peoples. The third and final dimension of global environmental discourses to be explored is how both images of difference and processes of hybridisation can work either to the benefit or detriment of indigenous and other marginalized peoples. There is a recurring pattern whereby erstwhile environmental 'villains' appropriate the concepts of alterity deployed against them and redeploy them so that they look instead like 'heroes'. We will illustrate this exertion of local agency with case studies of environmental learning among migrant *colonos* in the Brazilian Amazon, representation of ethnic harmony among Punan hunter-gatherers in Indonesian Borneo, and the self-proclaimed role of urban migrants in environmental preservation and restoration in Kathmandu. Alterity, hybridity and their strategic uses all reflect and contribute to the paradoxical relationship between Western and non-Western systems of environmental knowledge in the modern world.[7]

Western Discourses of Difference

The significance of the assumed difference between Western and non-Western societies is suggested by its persistence in diverse circumstances, often in the face of evidence of longstanding conceptual, social and material exchange between the two supposedly separate realms (Gupta and Ferguson 1997). In this section, we illustrate these tendencies with two seemingly disparate examples – one regarding Western images of swidden agriculture as both non-Western and environmentally destructive, the other regarding a case of regional conservation *within* the West where a perceived urban/rural divide closely resembles the West/non-West divide. Together, these examples show that the nature and location of such divides is defined more by political struggles than by any inherent social or cultural characteristics. They also illustrate how the perceived divide produces the symbolic resources that are crucial to struggles over identity and resources that are discussed in following sections.

The Exoticisation of Swidden Agriculture

A powerful symbol of Western views of non-Western environments is swidden agriculture, also known less accurately as 'shifting cultivation' or more deprecatingly as 'slash-and-burn agriculture'.[8] During the second half of the twentieth century, the image of a poor farmer standing in a swidden full of charred tree trunks became ubiquitous in Western representations of non-Western environmental degradation, especially tropical deforestation. This image explicitly stands for the purported pressure of poverty on the environment; more implicitly, it stands for the purportedly short-term, irrational and destructive use of natural resources by non-Western farmers.

The profoundly negative loading of this image is reflected in the role that the term 'slash-and-burn' has come to play in Western language. During the corporate downsizings of the late 1980s and 1990s in the United States, 'slash-and-burn' became a popular term for closings of plants and laying-off of workers (*Wall Street Journal* 1989, *Newsweek* 1993). The budget-cutting and elimination of government departments and programmes that came to characterise the Republican administrations of the 1980s and 1990s similarly came to be associated with the phrase 'slash-and-burn', giving us the phrases 'slash-and-burn budget', and 'slash-and-burn revolution'. Common to these uses of 'slash-and-burn' is the connotation of a contest waged with ruthlessness and broad savage strokes (albeit with the ostensible aim of producing a 'leaner' and more efficient organisation).

Over half a century of ethnographic research has clearly shown a picture of swidden agriculture quite unlike these popular representations. In most tropical forests, the nutrient stores that can be exploited by agriculture lie mostly not in the soil but in the biomass atop it, and it is this that swiddens exploit.[9] This is reflected in the term for swidden cultivation among Ibanic-speaking Dayak in Borneo, *bumai hutan* ('farming the forest'). The nutrients in the biomass are extracted through burning, which breaks down the biomass into a nutrient-rich ash that cultigens can easily access.

Swidden agriculture's reliance on natural forest dynamics to restore fertility after each cropping cycle gives it one of the greatest returns to labour (not land) known in agriculture. It has been calculated that the return on labour in swiddens is two to four times as great as that of irrigated rice terraces (Ruthenberg 1976; Dove 1985: 6). Proof of the economic and ecological sustainability of swidden agriculture lies in its sheer persistence. Recognising that global surveys are challenged by the sheer diversity of practices that might be labelled 'swidden cultivation', estimates nonetheless suggest that agricultural technologies that might be subsumed under this label are practised on 30 percent of the world's arable soils (Bandy et al. 1993: 2) and support as many as one billion people – 22 percent of the population of the developing world in tropical and sub-tropical zones (Thrupp et al. 1997: 1–4).

When Western scientists and government officials penned accounts of swidden agriculture in the tropics in the nineteenth and early twentieth centuries, they typically noted how alien it appeared to their eyes. This perceived alienness rests, however, on a forgetting of Western agricultural history, in particular the role of swiddens in Western Europe and North America. In the Ardennes in France, swidden cultivation was actually a more profitable occupation than paid labour in industry until early in the twentieth century (Sigaut 1979: 685). A swidden system based on a melding of Scots–Irish and Native American practices developed in the uplands of the southern United States, where it dominated through the nineteenth century and has persisted to the present day (Otto and Anderson: 1982).

A reason for forgetting this history is the critical attitude of the modern state and its representatives towards swidden cultivation. The tenacity of this official anti-swidden discourse is reflected in Kleinman et al.'s (1995: 235) comment on the '[continued] inability of domestic and international development agencies to consider slash-and-burn agriculture as a sound food production system'. The antipathy of the modern state to swidden agriculture is based on its 'illegibility.' States tend to favour intensive, fixed-field, infrastructure-heavy agriculture because its product is visible, concentrated and susceptible to state extraction, and its people are tied by capital investment to their fields. States tend not to favour swidden agriculture because, in contrast, its product is far less extractable and, in the absence of capital investment, its people are far more capable of evasion and, if necessary, flight.[10] As Scott (1998: 282) writes, swidden is an 'illegible form of agriculture', comprising 'fugitive' fields and cultivators, and constituting 'potentially seditious space'. The modern state's criticism of swidden agriculture is thus, in reality, a question not of agro-ecological development but of political-economic control.

The current antipathy in Western development circles towards swidden agriculture must be interpreted in light of the West's own swidden history. The erasure of this history and the rise (especially among bureaucrats and scientists) of a critical, anti-swidden discourse appear to have occurred in the developed Western nations precisely as the practice of swidden was waning there but continuing in, and becoming solely identified with, the less-developed non-Western nations. This coincidence suggests that Western deprecation of swidden agriculture is not so much a function of its geographic, historic and technological distance as it is part of a political effort to *make* it distant. This socially constructed alienness helps Western nations to adopt a critical, self-empowering view of natural resource use in developing, non-Western nations. Such discourses of 'under-development' help to legitimate intervention by the part of the world that wields them in the lives of the part that is caricatured by them (Ferguson 1990; Escobar 1995).

The Construction of the 'Northern Forest' in New England and New York

It is instructive to examine not only how difference is constructed between Western and non-Western societies, but how this process occurs in structurally similar situations within the West itself. The rhetoric used to promote conservation and development interventions in many domestic U.S. contexts draws heavily on essentialised images of rurality and wilderness and greatly resembles that employed regarding non-Western nations. Both narratives draw on an overarching discourse that portrays supposedly less modernised environments and associated peoples as fundamentally 'other'.

An example can be found in the Northern Forest of New England and New York, a region that was defined in 1988 and has, since then, become a focus of conservation in the northeastern U.S. (Klyza and Trombulak 1994). The socially constructed essence of the Northern Forest is suggested by a poster produced in 1998 by a coalition of environmental organisations (Northern Forest Alliance 1998). A photograph, looking down on the sunrise from a mountain summit, shows clouds below, dappled with shadows and shades of pink; an expanse of unbroken forest; and a remote pond tucked among the folds of a massive, fog-shrouded ridge. The poster invites us to 'Explore the Northern Forest,' which is identified on a map as a green mantle draped across the northern portions of Maine, New Hampshire, Vermont and New York. We are told to 'experience the landscape, the culture and the heritage of … the largest and last continuous wild forest east of the Mississippi River', with its high mountains, 'pristine lakes and rivers', and 'remote wetlands'. There are people here, too – mostly descendants of European immigrants who are said to have 'grown up hunting, fishing, trapping, and walking in the woods …'. Although we are told that some now work in business or manufacturing, they are 'proud of their heritage, and a way of life so different than in the urban areas around them'. Finally, we learn that there are problems here, like development of lake shores, reduced recreational access, and loss of jobs in the forest industry. It is said to be 'up to us' to save the Northern Forest, to leave, for our children and grandchildren, a healthy forest and strong communities that can continue to support a way of life that has existed for generations'.

This image draws on central themes that appear regularly in environmentalist literature. Of particular importance is the fundamental otherness of the Northern Forest, which constructs for visitors (primarily metropolitan residents) a symbolic and experiential opposite of the everyday and the mundane as the basic consumable resource of tourism (Urry 1990). The Northern Forest is depicted as immense, wild, natural and strikingly beautiful. It is represented as a contiguous, cohesive region that stands in opposition to the surrounding urban and suburban landscape so often lamented as artificial, confining, predictable and unattractive. It is a place that visitors can retreat to as a way of escaping the stresses of modern life and exploring new places and possibilities. Although local people are included in this vision, they, too, are very different from the 'us' that is implicitly the poster's main audience. People in the Northern Forest harken back to an idealised, imagined past and fit seamlessly into the natural landscape.

These images of difference have great appeal because they connect to popular narratives of the frontier as a land of new possibilities; of sublime nature as a means of escaping or transcending modernity; and of the fusion of both in the contemporary emphasis on wilderness as both recreational and spiritual retreat (Cronon 1995). This conception of the Northern Forest, however, is highly problematic, in large part because it erases the pervasive linkage and hybridisation of this supposedly non-modern landscape and nearby metropolitan areas. Far from being isolated in time and space, for example, this northern border country has been tightly bound to nearby urban centres for more than two centuries through social, cultural and economic ties, including a pervasive influence of absentee ownership of land and capital (Luloff and Nord 1993). Nor do the Northern Forest's boundaries fit neatly with biophysical or social indicators. The Appalachian Mountains and associated forest types bend southward through the length of New Hampshire and Vermont, while Lake Champlain and extensive farmland in northwestern Vermont create a sharp break between its eastern and western sections.

On large corporate holdings within the Northern Forest, intensive forest practices and a dense network of logging roads have caused dramatic ecological changes, especially since the 1970s. Life in these communities has changed markedly in the past several decades. There has been mechanisation and job loss in the forest industry and a consequent increasing dependence on tourism, government and service industries. With improved transportation has come the growth of regional commercial centres, decreased rural isolation, and a loss of community cohesion and local economic activity. There has been a pronounced shift of political authority from local to state and federal governments (Hays 1987). Demographic changes include a long history of out-migration of youth seeking economic opportunity and more recent in-migration of people seeking recreational amenities and a rural lifestyle. Finally, the very concept of an interstate region called the Northern Forest did not exist until 1988, when the sale of large tracts of industrial forest land in each of the four states drew the attention of both government and environmental groups. Only then was the region named and institutionalised for the purposes of policy initiatives and political advocacy (Reidel 1994).

As with the previous example of swidden agriculture, the successful circulation of images that are so easily questioned suggests that they are not so much 'mistaken' as ordered and power-laden constructions of knowledge (Ferguson 1990). While there are, indeed, significant commonalities across many parts of the Northern Forest – mountains, infertile soils, low population density, recreational tourism centres, large private landholdings and industrial forestry – its status as a 'region' was hardly inevitable. Rather, its construction rests on deep-seated metropolitan conceptions of rurality and wilderness that have been further promoted by non-local environmental groups in their appeals for financial and political support (Smith 2003).

For urban environmentalists and recreational tourists, the Northern Forest is principally a place to visit and to explore interesting natures and cultures. Whatever benefits or drawbacks may emerge for local residents, the Northern Forest remains in many ways a place to be controlled, utilised, and conserved by non-residents. This example illustrates how marginalised societies and environments in developed countries may be defined and objectified in the same ways as their counterparts in less-developed countries. The conceptual and political dynamics at play in both scenarios are borne more of political-economic *relationships* between core and periphery – relationships that, crucially, occur at both national and global scales – than of imagined geographic *divides* between country and city or between East and West.

Hybrid Systems of Knowledge

A prominent feature of global environmentalism since the 1970s has been the discourse of indigenous environmentalism, in which indigenous peoples are portrayed as protecting nature due to their deep ecological knowledge and cosmologies that stress a kinship with, rather than dominance of, the natural world. In this same discourse, Western science is often posed as a polar opposite to indigenous knowledge, objectifying nature in order to manipulate it. Whereas this represented a welcome corrective to a century and more of virtual denial by the West of the existence of indigenous knowledge in non-Western regions, it nonetheless represents a simplistic understanding of the history of difference between Western and non-Western knowledge. Here we present two cases in diverse settings (Mexico and Indonesia) that make clear that knowledge systems of many indigenous peoples are actually complex hybrids resulting from long-term historical interactions.

A Hybrid Forestry System in the Sierra Juarez

The historical hybridity of knowledge systems is well illustrated by the forestry practices of the Zapotec communities of the Sierra Juarez of Oaxaca, Mexico. These communities have been widely praised for their careful and sustainable forest management (Bray 1991). Advocates of indigenous knowledge attribute this success to an ethic of forest protection, based on traditional agro-ecological knowledge. Tyrtania (1992), among others, has documented the impressive complexity of traditional resource-use systems among the Sierra Zapotec. And, indeed, throughout Mexico, as elsewhere, scholars and politicians in recent years have credited indigenous peoples with extensive ecological knowledge and the use of sophisticated techniques of forest management. However, crediting the Zapotec forestry successes to strictly local knowledge represents a denial of history.

In the 1930s the Mexican Forest Service, influenced by contemporary U.S. Forest Service policies (Simonian 1995), imposed upon Zapotec communities the duty to form fire-fighting brigades and to suppress fires. The incidence of fires did not markedly decline until the late 1940s, however, when commercial logging began in the forests of Ixtlán. An outside company employed *comuneros* – villagers with property rights – as loggers, providing them with cash income and an alternative to subsistence cultivation. From 1956 onwards, the view that fire was destructive was further strengthened by the actions of the forest concessionaire company FAPATUX, which built fire towers and organised fire brigades. In 1982, the community took responsibility for managing its own forests, largely continuing the fire management practices it had inherited from FAPATUX. Although in traditional Zapotec swidden agriculture (*milpa*) fire was an important tool and was also used to encourage pasture growth for cattle and sheep, today residents of Ixtlán involved in logging describe fire fighting as a shared obligation and fire as a destructive agent, let loose by malicious, stupid, or careless people, principally from neighbouring communities (Mathews 2004).[11]

Successful forest protection in Ixtlán is based not only on the impact of modern scientific ideas about the forest but also on the community's success in incorporating forestry science into community management structures and practices, the latter of which are the result of a profound reordering of community life during the colonial and post-colonial periods (Chance and Taylor 1985; Chance 1989) and so are not in any simple sense 'non-Western' although they contain elements of pre-colonial political traditions. A critical factor in contemporary forest conservation in Ixtlán de Juarez is its large area of forest (19,000 ha, for a relatively small population of 2,100), which reflects its successful manoeuvering during the political struggles of nineteenth- and twentieth-century Mexico. Ixtlán was a military supporter of presidents Benito Juarez and Porfirio Díaz and later of the ultimate winners of the Mexican revolution. In the nineteenth century Ixtlán was one of the few communities to successfully petition to have its boundaries surveyed, and it has since been able to retain much of its large landholdings in spite of attempts by sub-communities to break away (Garner 1988; Pérez García 1996 [1956]). More recently, Ixtlán has been selected as the location for a new secondary school and government offices, thereby creating a large pool of *comuneros* who are trained in forestry and accounting and affording them continued opportunities to learn how to manipulate government bureaucracies. At the same time, the community has been able to hold unwanted government bureaucracy at a distance; a recent government land-titling project was only allowed to survey the external boundaries of the community, with internal boundaries being regarded as community business.

In Ixtlán, *comuneros* have been able to incorporate modern forest science into community forest management through a mixture of pragmatism and political guile. They do not hold a unified non-Western cosmology or science, proffering instead local explanations of specific practices and blending modern scientific forestry with traditional agricultural practices. In spite of negative impacts from logging in the past, the community of Ixtlán has been able to combine modern forest management with its own traditional political organisation to protect the forests of the Sierra Juarez.

Non-Western Uses of Mapping Technology

Another way that non-Western systems of environmental knowledge hybridise with science is through the adoption of Western methodologies for *representing* that knowledge, one of the earliest (and still most important) of which is mapping. Both descriptive and normative, maps can reflect political boundaries, natural features, local resources, social structure, or cosmology (Brody 1982; Thongchai 1994; Peluso 1995; Mundy 1996; Scott 1998). Part of the power of maps in the modern era has derived from their supposed neutral and objective representation, as typically employed by ruling powers in the West (Kain and Baigent 1992; Riles 2004). This has given rise over the past decade or so to a critique and a counter-mapping movement among scholars and activists in non-Western nations.

Before the colonial era, indigenous maps existed in some regions but were very sparse in others. China, Vietnam and Burma have rich cartographic traditions, whereas there is virtually no record of precolonial maps for Cambodia, Laos and insular Southeast Asia (Schwartzberg 1994). With the spread of colonialism and the concomitant European effort to extend control over land, population, and production in Southeast Asia, maps took on a central importance for new rulers, with increased attention to detail and frequent revisions. Colonial use of mapping shifted from the early exploration of territory to efforts increasingly focused on delineating the boundaries of administrative units (Thongchai 1994; Henley 1995). There were notable differences among the colonial powers in their attitudes toward traditional land claims and uses, and these differences are reflected in both mapped representations and in colonial land regulations. Towards the end of the colonial era, mapping also played a crucial role in shaping national and regional identities (Anderson 1991; Thongchai 1994).

Modern states, and some conservation organisations as well, still use maps to limit local residents' land claims by defining such use as 'encroachment' in conservation areas or in regions designated for some other form of development. For example, representing forests used by swidden cultivators as 'empty' on official maps is a strategic move which ignores people's presence and denies the legitimacy of their land use, thereby bolstering the case for alternative claims (Li

1996). However, local people also have begun to employ the Western technology of boundary mapping to communicate their resource claims in a way acceptable to modern national authorities (Tsing 1999, Zerner 2003). Community-level mapping, which was developed and popularised primarily in South and Southeast Asia (Chambers 1997), uses participatory methodologies to reverse the flow of information from externally produced to locally informed maps. This is part of a wider-ranging critique of the Western scientific discourse of mapping as a state-directed activity that produces authoritative documents (Scott 1998; Harwell 2000), in which many map makers now define their goals as the promotion of community interests (Sirait et al. 1994). Community mapping is also vulnerable to some of the constraints of the medium, however, including the difficulty of reflecting the dynamic nature of traditional land claims and use alongside other land designations (Fox 1998). Other concerns include the implications of fixing ethnicity to defined spaces and the difficulty of representing complex local land classifications (Tsing 1999; Li 2000).

Evidence from an interior valley in the Bird's Head region of Papua, Indonesia, illustrates the shifting purposes and multiple actors involved in mapping today. This region is inhabited by tribal clans who practise swidden agriculture alongside hunting and gathering in the forest. In 1999 the most topographically accurate maps available were Dutch maps from 1957, which were based on 1944 aerial photographs taken by the United States military. The Dutch used these maps in their regional planning for forestry development and plywood production in the valley. Subsequent Indonesian government maps drew on the information in the Dutch maps with varying degrees of precision, but to this day many official maps of the region inaccurately reflect the human settlement in the valley. In addition, a review of the maps in use by different government departments under the pre-1998 New Order regime showed the region variously categorised as targeted for park protection, forestry development, transmigration, or conversion to agro-industry.

Villagers, in partnership with a local legal aid society and the agriculture faculty of the provincial university, undertook a valley-wide mapping effort in the late 1990s to represent their forest practices and understanding of traditional clan boundaries to Indonesian government officials. They used the 1957 Dutch topographical maps as a reference, supplemented by participatory mapping techniques, with a plan to combine all of the data gathered using geographic information systems. Non-Western communities have thus taken up the Western concept of mapping and transformed it into a tool with which to challenge official views by presenting alternative views of resources and territories. As with the Zapotec, practical knowledge in this case is fundamentally both hybridised and political.

Heroes and Villains in Resource Conservation

Images of alterity supply the symbolic resources that are used by various groups in struggles over identity, authority, and natural resources. Once they are conceived as fundamentally 'other', indigenous and local communities may be objectified *either* as heroic champions of the environment *or* as villains of resource destruction. But crucially, strategic efforts by marginalised groups to make use of positive images of indigeneity for their own benefit are, almost by definition, markers of hybridised knowledge systems; they must make use of Western notions of difference in order to assert their indigenous, non-Western identity.

Nowhere are these tendencies and their political implications better illustrated than in tropical rainforests. Beginning in the 1980s, many Western environmentalists concerned with conserving tropical ecosystems championed their cause with romanticised representations of forest-dwellers as 'ecologically noble savages' or primitive environmentalists. These representations essentialise and exoticise forest-dwellers as timeless, egalitarian, wise, and natural stewards of the environment (e.g., Lynch and Talbott 1995). Some anthropological work criticises these representations of forest-dependent communities, revealing their dubious authenticity, the Western environmentalist agendas that motivate them, and the political-economic consequences when one group of forest dwellers versus another captures the spotlight of Western environmentalist interest (Ellen 1986; Brosius 1997; Conklin 1997; Li 1999). Other scholarship has taken a different tack, examining not only the process of fashioning these representations, but also the benefits they provide to local people (Tsing 1999; Li 2000). This section presents two cases, from Brazil and Indonesia, that show how local communities in forested regions of these two countries have used essentialised representations to their own advantage. A third example, involving squatters in Kathmandu, exhibits very similar dynamics in an urban setting and shows that the basic contours of this phenomenon are recurrent in diverse contexts.

Migrant Farmers on the Amazonian Frontier

Beginning in the 1970s, both the indigenous tribal peoples and the long-settled rubber tappers of the Amazon gained recognition as 'natural conservationists' (Conklin and Graham 1995). The high-profile nature of struggles for land rights by these 'green' actors has tended, however, to obscure the fate of many of the other people in the region, including the far more numerous migrant farmers. In sharp contrast to the 'green' image of indigenous forest peoples, small migrant farmers in the Amazon have been viewed as the 'villains' of the forest, and they have been ignored by researchers and officials as potential allies in forest conservation. There are two reasons why migrant farmers have rarely been thought of as potential allies in forest conservation: first, they are not 'native' and so are not thought to have any locally specific knowledge about the appropriate use and management of the forest; and second, the logic behind their existing use of the forest has been neither adequately nor sympathetically examined.

The small migrant farmers (locally known as *colonos*) who populate the Amazonian frontier hail from southern and northeastern Brazil and are a heterogeneous group (Moran 1981). Their migration to the frontier has been a response to a variety of political and economic forces, including land availability, financial incentives through government programmes, massive road building, economic opportunities, and economic failure in their place of origin (Hecht and Cockburn 1989; Schmink and Wood 1992; Hall 1997). They have played an important symbolic role at the frontier for the Brazilian government, which has been able successfully to characterise them as the 'villains' of Amazonian deforestation. By blaming the migrants for the ecologically disastrous consequences of its own development programmes, the government has been able to deflect most of the blame from itself and the equally culpable private sector.

In the absence of incentives and infrastructure to encourage sustainable land use, poor *colonos* are in fact often obliged to continue to mine the natural resource base rather than make long-term investments in it (Pichón 1996). However, agriculture intensification and a shift towards fixed-field systems do begin to make economic sense in older frontier areas, where there is greater access to both markets and technology, and the scarcity of land and rising land prices make swidden agriculture and cattle ranching less viable. In these areas, colonists who arrived not even two decades earlier have begun to experiment with natural forest management and agro-forestry, partially reconstructing the tropical forest on their farm plots. These non-indigenous farmers have started to develop their own systems of environmental knowledge, drawing on their background, culture and society; on their experiences at the frontier; and on knowledge acquired from other groups there (Moran 1981, Hall 1997). At the same time, these older migrants have started to organise themselves to fight for political legitimacy and recognition for their resource-use systems.

Recent grassroots initiatives show the increasingly active stance being taken by migrants. For example, the *Movimento pelo Desenvolvimento da Transamazônica e do Xingu* or MDTX ('Movement for Development on the Transamazon and Xingu regions'), developed in association with the Catholic Church, is an umbrella institution for forty local organisations including rural unions, farmers' cooperatives, teachers' and health workers' organisations, and movements of women, youths, and blacks (Hall 1997). The MDTX's first major initiative, pursued in concert with a number of other grassroots institutions, was to propose a development programme in which government and farmers' organisations would collaborate to reconcile the twin objectives of environmental conservation and smallholder agricultural production. The proposed programme consists of plans to: (1) reorganise land tenure throughout the region; (2) disseminate sustainable agro-ecological technologies; and (3) establish major new conservation areas in the region. The second notable grassroots initiative by migrants involves a proposal to reformulate the government agricultural credit line known as *Pro-Ambiente* (Pro-Environment). This new environmental credit arrangement would provide incentives for sustainable production systems and extractive activities by compensating farmers for the expenses that they incur to protect natural watercourses, shift to permanent forms of agricultural production, and re-establish forest on cleared land that is not suitable for agriculture.

These grassroots initiatives reflect the political astuteness of the migrants in trying to assume the newly powerful role of environmental steward. More broadly, they reflect the migrants' concern to be represented neither as the 'villains' nor as the 'victims' of development in the region but to structure for themselves their own, more active role in development.

Primitive Conservationists in Borneo

Another example of the way that local communities are responding to the preconceptions of Western environmentalists comes from East Kalimantan, Indonesia, where the Center for International Forestry Research (CIFOR) attempts to develop forest co-management through applied research with local Dayak communities and other stake holders, thereby opening up new space for Dayak to contest dominant state resource discourses and to better represent their own resource rights and uses.

The implications were demonstrated during a two-week research workshop held by CIFOR in 1999, which was attended by three local Dayaks. Each of these individuals represented a different ethnic group – Kenyah, Merap and Punan. All three men were well-respected in their communities and had extensive prior contacts with researchers. During the workshop, they all emphasised both the cohesion among their ethnic groups and their joint efforts to protect forest resources. For example, they explained that when they find gaharu (*Aquilaria* spp. – exported for use in perfumes and incense), they attempt to extract the part of the tree that is infected and therefore valuable, leaving the rest of the tree standing in the hope that it will recover. This account contradicts previous descriptions by other Dayak of the gaharu harvesting process, in which the entire tree is felled. Further, the three workshop attendees consistently spoke of the harmonious cooperation among their respective ethnic groups, presenting a picture of village social dynamics which dramatically differed from both the stories told previously by other villagers and direct observations during fieldwork in their villages. This fieldwork revealed not only a lack of cohesion among the different ethnic groups, but also an explicit inter-group discourse of inferiority and aggression. For example, Kenyah villagers frequently refer to Punan and Merap in disparaging terms (e.g., as *terasing* [isolated] and *terbelakang* [backwards]), often deploying the same rhetoric used by outsiders when referring to Dayak generally. Merap and Punan, meanwhile, often comment that Kenyah are aggressive land grabbers, and selfish.

These discrepancies are explained by the shift in sociopolitical context between village and research station. At the research station – which represented, in comparison with the village, a shift to the formal and public – these Dayak men were engaged in an event of formal documentation for the international research organisation. It represented what Tsing (1999) calls a 'field of attraction', referring to the emergence of a setting that 'nurtures and maintains the relationship between the rural community and its [environmentalist and green development] experts … that keeps experts coming back'. The Dayak were, in short, representing themselves to CIFOR as the kind of people with whom they thought such an organisation – one focused on sustainable forest management and improved local livelihoods – would want to work. This is not necessarily to imply that these representations were convincing to CIFOR researchers, but rather to note that it was an attempt by villagers to articulate and align themselves with the organisation, which they perceived as possibly bolstering their politically weak position (Li 2000).

Another example of Dayak adoption of Western environmentalist discourse involves the articulation of resource rights claims in romanticised terms. In 1991, a number of Punan in this same region of East Kalimantan formed a foundation to serve as the 'official' voice for Punan living throughout the province. One of the documents produced by the foundation, entitled 'The Mission and Vision of the Community of Traditional Punan Dayak', exemplifies the use of metaphors of nature and ancient customary law that appeal to Western environmentalist fantasies of forest dwellers:

> The words of our Mission and Vision … are as strong as ironwood and as hard as iron stone. The wisdom of customary law for the Punan community … is not merely a theory or concept like the products of laws, presidential decrees and government regulations that … have caused losses in the rights of customary communities. Anybody … who does not respect the existence of custom means that he or she is not the creation of God, who said humans must live and have children and grandchildren just like the grass and wood that is above the earth, guard and be responsible for the protection of nature in the whole world.

Even in less formal contexts, Punan in this region now invoke images of themselves that fit well with Western environmentalists' notions of the indigenous naturalist. For example, Punan who have worked closely with both researchers and NGOs (non-governmental organizations) may use expressions such as, 'The forest is to us as milk is to a baby', which is a non-traditional image. And they frequently mention the abundance of medicinal plants in the area and their knowledge of them. This echoes Brosius's (1997: 62) account of Penan in Sarawak who had worked with environmental NGOs: 'One of the more interesting consequences of the environmentalist rhetoric of medicinal plants is that this rhetoric has itself suffused back to the Penan and been adopted by them as their own.' The purpose of these observations is not to evaluate the degree of authenticity of the Western environmentalist representation adopted by these forest communities, but to illustrate the way that the articulation (Li 2000) of these Western representations can give political influence to such communities.

Ecological Restoration in Kathmandu

Rhetoric traceable to Western institutions can be found not only among those with power to locally frame ecological issues, but also among those who might contest this framing. This can be seen in Kathmandu, which over the last decade has assumed a prominent place among South Asia's fastest-growing cities.[12] A contemporary urban environmental issue is the plight of the urban reaches of the Bagmati and Bishnumati Rivers, which converge in the city. The rivers suffer severe ecological degradation, characterised by extremely poor water quality, serious morphological changes, and, some argue, loss of the cultural and religious values traditionally attributed to the rivers. Comprehensive policy and development studies identify the main causes of river pollution inside the urban area as the discharge of untreated sewage and widespread dumping of solid waste into the rivers and on their banks. Excessive sand mining in river beds and banks, which supplies building materials to the city's construction industry, is blamed for significant morphological change and severely channelised flow patterns.

In addition, most discussions of river degradation identify human encroachment on the banks, floodplains, and river beds as a significant factor in the process. For migrants from Nepal's countryside as well as for poorer city residents, participation in the legal land and housing markets is impossible. As a result, many of these new residents, as well as long-term Kathmandu residents, have joined the swelling numbers of *sukumbaasi* (squatter) settlements along the rivers. In 1991, these settlements were estimated to be growing by 12 percent annually, a rate twice that of the city itself. By 2000, the growth rate had slowed, but a significant portion of the urban riparian corridor is lined with the semi-permanent structures of settlers.[13] Implicit in many government plans for riparian restoration is the forced removal of existing *sukumbaasi* settlements.

Advocates of settlement rights for riparian *sukumbaasis* are trying to counter the government's representation of their role in riverian degradation through a counter-narrative with clear ties to Western ecological rhetoric. By emphasising Western development concepts such as 'healthy cities' and 'sustainable human settlements', settler advocates offer a narrative that inserts socioeconomic concerns into ideas about the ecology of the rivers and the city itself. Although they may not directly contest the official state narrative of river degradation, squatter advocates are using their own notion of ecology to argue for precisely the opposite of the fate for riparian settlements called for in the official restoration scenario: they argue that *upgrading* squatter settlements, rather than *removing* them, is the key to realising an ecologically healthy riverscape. The squatter advocates' rhetorical strategies can be traced to the United Nations-sponsored *Future Cities*

World Habitat Day Conference (FCWHD), held in Kathmandu in 1997. The emphasis of the conference was on blaming not squatters adjacent to the rivers but insufficient infrastructure throughout the city for river pollution. Rather than being seen as the cause of river degradation, *sukumbaasis* were discussed as its *victims*. By invoking a UN 'healthy cities' model, conference presenters constructed an urban ecology in which a 'healthy environment' is assessed through its capacity to provide the essentials of life to its human inhabitants.

Advocates for the squatters resisted any discussion during the conference of the negative effects riparian settlements might have on the ecological integrity of the rivers. Whereas they agreed that proximity to the degraded resource *seemed* to implicate settlers in the degradation process, they argued that in ecological terms *sukumbaasis* play only a minor role. This is most obvious with sewage. Since Kathmandu lacks a comprehensive, functional sewage treatment system, effluent inputs originate throughout the city, implicating squatters and non-squatters alike. Squatter advocates further argued that since *sukumbaasis* consume fewer goods than more wealthy urban inhabitants, their per capita contribution of the by-products of industrial production to the river system is probably relatively small.

At the local level, many *sukumbaasi* communities are in fact actively engaged, either collectively or individually, in activities that contribute to the monitoring and/or improvement of river quality. For example, settlers patrol their settlements for illegal riverside dumping, practised widely by the city municipalities, and plead for more authority to watch for, and halt, solid waste dumping on the river banks. Planting vegetation on the river banks is also a common practice among the squatters (ironically so, since riparian re-vegetation is a goal of the anti-squatter state development narrative). The squatter perspective on urban ecology not only downplays any deleterious effects the settlements might have, it constructs *sukumbaasis* as in many ways more ecologically minded than their non-squatter neighbours. It directly contradicts representations of *sukumbaasi* knowledge, attitudes and practices included in official characterisations of river pollution (Rademacher 1998) . By claiming rhetorical tools and sentiments from a Western development arena, *sukumbaasi* advocates have successfully countered a state development narrative on its own terms, contesting not only a particular environmental programme but the very definition of 'ecology' itself.

Conclusions

We have examined here the global mobilisation of environmental concepts and its implications for the perceived divide between Western and non-Western systems of environmental knowledge. The recent re-evaluation of indigenous knowledge has heightened popular belief in a Western/non-Western divide with respect to sustainable versus unsustainable environmental relations. But the case studies of environmental knowledge presented here clearly show that this divide is socially constructed. Western and non-Western traditions are increasingly intermingled and both also encompass multiple, diverse, and sometimes conflicting paradigms. What appear to be non-Western systems of environmental knowledge typically prove to have been hybridised with or otherwise related to Western systems of knowledge. In the face of this history of East–West linkages, the popular currency of a divide testifies to its strategic utility. Essentialised images of non-Western resource users are powerful political symbols, which are adopted and used by Western and non-Western actors alike.

Our findings regarding the global mobilisation of environmental knowledge, and the constructed character of the divide between Western and non-Western environmental knowledge, raises the question of why belief in this divide is currently flourishing among practitioners and in popular consciousness. Why has the project of differentiation between East and West emerged (or re-emerged) at this time in global history? Why is the distinction between East and West being reified precisely at the point in history when it seems to be losing whatever empirical validity it may ever have had? The answer, we suggest, has to do in part with the utility of this distinction for societies trying to protect themselves against some of the costs of political and economic globalisation while at the same time availing themselves of some of the benefits.[14] Why is the environment so central to this utility? Why, notwithstanding the fact that much knowledge of the environment is globally generated and mobilised today, has such knowledge nonetheless emerged as a key domain for global differentiation? Why are both Western and non-Western actors alike using the environment as a focal point for reiterating the myth of themselves in contradistinction to one another? We suggest that the answer to this second set of questions involves, in part, the place-based connotations of environment and the importance of place to social identity and the capacity for political mobilisation in the face of the challenges of globalisation.

This study also raises a related set of questions about the proper role for anthropology to play in current debates about Western versus non-Western environmental knowledge. Anthropologists are not only concerned about the possibly negative implications of assertions of indigeneity, as noted earlier, they are also concerned about the implications of their analysis of such assertions. The appropriate relationship among analysis, advocacy and activism, has drawn the attention of a number of anthropologists (e.g., Brosius 1999; Hodgson 2002). This relationship was one of the issues raised in the recent debate about Kuper's (2003) critique of the concept of indigeneity and the possible political ramifications of this critique. Kuper dismisses such concerns, saying that it is 'our business ... to deliver accurate accounts of social processes' regardless of any 'fear of undermining myths of autochthony'. This stance seems to assume a level playing field. But as Hodgson (2002) points out, anthropology cannot be equally accountable to all, because all are not equally powerful. Kuper's critique would take away all of the tools of culture, race, history etc. by which unequal political-economic power can be contested. As Robbins (commenting in Kuper 2003) argues, deconstructing essentialist ideologies should be only the first step of anthropological practice, not the end goal, and denying access to resources to redress injustice because they deploy essentialist ideologies of culture and identity is a disservice to anthropology and activism.

The underlying issue here is the basic relationship of science to society in the modern global community. What do we make of the extraordinary coincidence that anthropology (and the other social sciences) began to critique the concept of indigeneity at the very time that it was being legitimised by mainstream global organisations like the United Nations and the International Labor Organization? How is academia's assault on locality related to the larger assault on locality of modernity? The paradoxes of modernity are not merely good subjects for academic study, they are quite likely to be processes in which academia itself is heavily imbricated.

Notes

1. This reversal was reflected in efforts by the United Nations (in 1986) and the International Labor Organization (in 1989) to formally define and protect indigenous rights and by the emergence of indigenous environmental knowledge as a specific sub-field of study and practice.

2. This is part of a wider, anti-essentialist trend in anthropology (Vayda 1990).

3. For Li among others, one of the inherent problems with reliance by marginal groups on the concept of indigeneity to empower themselves is the potential for indigenous status to be challenged. If you can be indigenous, then you also can be – or become – non-indigenous; and whereas the loss of indigeneity is seen as 'natural', its conscious construction is not.

4. Scott (1998: 331) writes: 'High modernism needs this "other", this dark twin, in order to rhetorically present itself as the antidote to backwardness.'

5. In the absence of such histories, Li says (2000), identity is 'fuzzy'.

6. Similarly, Li (2000) speaks of the 'places of recognition' upon which successful claims to indigenous status depend. In the absence of such places, even the seemingly most genuine indigenous claims will fail.

7. Parts of this analysis draw on Dove et al. (2003).

8. Most anthropologists abjure the use of the term 'slash-and-burn agriculture' as well as 'shifting cultivation', preferring instead to use the term 'swidden agriculture'. This is based on an archaic variant of Old English 'swithen', meaning to singe, which was resurrected because no contemporary term was sufficiently neutral to be used or even rehabilitated. Swithen/swidden was so archaic and indeed unknown as to have no negative connotations.

9. This is not invariably the case, however, as there are tropical regions with better soils and, consequently, more of a management focus on them (cf. Sillitoe 1996 on New Guinea).

10. The Southeast Asian literature, for example, documents instances of such flight in response to oppression by native states and colonial authorities and also as a response to contemporary periods of unrest.

11. The policy of fire suppression has dissenters within the community; many citizens in Ixtlán are not *comuneros*, do not benefit from logging, and would like to continue to farm using fire. One *non-comunero* criticised the community's stand against clearing and burning new swiddens, pointing out that pine trees came up spontaneously on old swidden fields. In fact,

almost everyone in the community is aware that pines naturally regenerate both on old forest fire sites and the sites of old swidden fields.

12. According to the World Resources Institute (1996), the urban growth rate in the cities of the Kathmandu Valley was 7.1 percent over the period 1990–95.

13. In the autumn of 1997, the total number of settlements characterised as *sukumbaasi* in Kathmandu was 54. Half of these were riparian – situated on the banks of the Bishnumati, Bagmati, or one of their larger urban tributaries (Tanaka, 1997.). Of the total population of *sukumbaasis* in the Kathmandu Valley in 1996 – close to 9,000 – 69 percent lived in riparian zones and about two-thirds of those occupied settlements on the banks of the Bishnumati or Bagmati Rivers.

14. We are indebted to Paul Sillitoe for his insights into this question (cf. Sillitoe 2002).

References

Agrawal, A. 1995. 'Dismantling the Divide between Indigenous and Scientific Knowledge', *Development and Change* 26: 413–39.

Anderson, B. 1991. 'Census, Map, Museum', in *Imagined Communities: Reflections on the Origin and Spread of Nationalism*, pp. 163–85. London: Verso.

Appadurai, A. 1996. *Modernity at Large*. Minneapolis: University of Minnesota Press.

Bandy, D.E., D.P. Garrity and P.A. Sanchez 1993. 'The Worldwide Problem of Slash-and-burn Agriculture', *Agroforestry Today* 5(3): 1–6.

Benjamin, G. 2002. 'On Being Tribal in the Malay World', in *Tribal Communities in the Malay World – Historical, Cultural, and Social Perspectives*, ed, G. Benjamin and C. Chou, pp. 7–76. Leiden: IIAS, Singapore: ISAS.

Bray, D.B. 1991. 'The Struggle for the Forest: Conservation and Development in the Sierra Juarez', *Grassroots Development* 15: 13–25.

Brody, H. 1982. *Maps and Dreams*. New York: Pantheon Books.

Brosius, J.P. 1997. 'Endangered Forest, Endangered People: Environmentalist Representations of Indigenous Knowledge', *Human Ecology* 25(1): 47–69.

———— 1999. 'Locations and Representations: Writing in the Political Present in Sarawak, East Malaysia', *Identities* 6(2–3): 345–86.

Chambers, R. 1997. *Whose Reality Counts? Putting the Last First*. London: ITDG Publishing.

Chance, J.K. 1989. *Conquest of the Sierra: Spaniards and Indians in Colonial Oaxaca*. Norman: University of Oklahoma Press.

Chance, J.K. and W.B. Taylor 1985. 'Cofradías and Cargos: an Historical Perspective on the Mesoamerican Civil-religious Hierarchy', *American Ethnologist* 12(1): 1–26.

Conklin, B.A. 1997. 'Body Paint, Feathers, and VCRs: Aesthetics and Authenticity in Amazonian Activism', *American Ethnologist* 24(4): 711–37.

Conklin, B.A. and B.A. Graham 1995. 'The Shifting Middle Ground: Amazonian Indians and Eco-politics', *American Anthropologist* 97(4): 695–710.

Cronon, W. 1995. 'The Trouble with Wilderness; or, Getting Back to the Wrong Nature', in *Uncommon Ground*, (ed.) W. Cronon, pp. 69–90. New York: W.W. Norton.

Dobbs, D. and R. Ober 1995. *The Northern Forest*. White River Junction, VT: Chelsea Green.

Dove, M.R. 1985. 'The Agroecological Mythology of the Javanese, and the Political-economy of Indonesia', *Indonesia* 39: 1–36.

——— 2000. 'The Life-cycle of Indigenous Knowledge, and the Case of Natural Rubber Production', in *Indigenous Environmental Knowledge and Its Transformations*, ed, R.F. Ellen, A. Bicker and P. Parkes, pp. 213–51. Amsterdam: Harwood.

Dove, M.R., M. Campos, A. Mathews, L. Meitzner, A. Rademacher, S. Rhee and D. Smith 2003. 'The Globalisation of Environmental Concepts: Rethinking the Western/Non-Western Divide', in *Nature Across Culture: Non-Western Views of the Environment and Nature*, (ed.) H. Selin. pp. 19–46. Dordrecht: Kluwer.

Ellen, R.F. 1986. 'What Black Elk Left Unsaid', *Anthropology Today* 2(6): 8–12.

——— 1999. 'Forest Knowledge, Forest Transformation: Political Contingency, Historical Ecology and the Renegotiation of Nature in Central Seram', in *Transforming the Indonesian Uplands*, (ed.) T.M. Li, pp. 131–57. London: Routledge.

Escobar, A. 1995. *Encountering Development: the Making and Unmaking of the Third World*. Princeton: Princeton University Press.

Ferguson, J. 1990. *The Anti-politics Machine: 'Development,' Depoliticization, and Bureaucratic Power in Lesotho*. Cambridge: Cambridge University Press.

Fox, J. 1998. 'Mapping the Commons: the Social Context of Spatial Information Technologies', *Common Property Resource Digest* 45: 1–4.

Garner, P. 1988. *La revolución en provincia: soberanía estatal y caudillismo en las montañas de Oaxaca (1910–1920)*. Oaxaca: Fondo de Cultura Económica.

Giddens, A. 1984. *The Constitution of Society*. Berkeley: University of California Press.

Gupta, A. 1998. *Postcolonial Developments: Agriculture in the Making of Modern India*. Durham, NC: Duke University Press.

Gupta A. and J. Ferguson 1997. 'Beyond "Culture": Space, Identity and the Politics of Difference', in *Culture, Power, Place: Explorations in Critical Anthropology*, (ed.) A. Gupta and J. Ferguson. Durham, NC: Duke University Press.

Hall, A. 1997. *Sustaining Amazonia*. New York: Manchester University Press.

Harwell, E. 2000. 'Remote Sensibilities: Discourses of Technology and the Making of Indonesia's Natural Disaster', *Development and Change* 31: 307–40.

Hays, S.P. 1987. *Beauty, Health, and Permanence: Environmental Politics in the United States,* 1955–1985. New York: Cambridge University Press.

Hecht, S. and A. Cockburn 1989. *The Fate of the Forest: Developers, Destroyers, and Defenders of the Amazon.* New York: Verso.

Henley, D. 1995. 'Minahasa Mapped: Illustrated Notes on Cartography and History in Minahasa, 1512–1942', in *Minahasa Past and Present: Tradition and Transition in an Outer Island Region of Indonesia,* (ed.) R. Schefold, pp. 32–57. Leiden: Research School CNWS.

Hirtz, F. 2003. 'It Takes Modern Means to be Traditional: On Recognizing Indigenous Cultural Communities in the Philippines', *Development and Change* 34(5): 887–914.

Hodgson, D.L. 2002. 'Introduction; Comparative Perspectives on the Indigenous Rights Movement in Africa and the Americas', *American Anthropologist* 104(4): 1037–49.

Hornborg, A. 1996. 'Ecology as Semiotics: Outlines of a Contextualist Paradigm for Human Ecology', in *Nature and Society: Anthropological Perspectives,* ed, P. Descola and G. Pálsson, pp. 45–62. London: Routledge.

Kain, R.J.P. and E. Baigent 1992. *The Cadastral Map in the Service of the State: a History of Property Mapping.* Chicago, IL: University of Chicago.

Kleinman, P.J.A., D. Pimentel and R.B. Bryant 1995. 'The Ecological Sustainability of Slash-and-burn Agriculture', *Agriculture, Ecosystems and Environment* 52: 235–49.

Klyza, C. McG. and S.C. Trombulak (ed) 1994. *The Future of the Northern Forest.* Hanover, NH: University Press of New England.

Kuper, A. 2003. 'The Return of the Native', *Current Anthropology* 44: 389–402.

Li, T.M. 1996. 'Images of Community: Discourse and Strategy in Property Relations', *Development and Change* 27: 501–27.

——— 1999. 'Marginality, Power and Production: Analyzing Upland Transformations', in *Transforming the Indonesian Uplands,* (ed.) T.M. Li, pp. 1–44. Amsterdam: Harwood.

——— 2000. 'Articulating Indigenous Identity in Indonesia: Resource Politics and the Tribal Slot', *Comparative Studies in Society and History* 42(1): 149–79.

Luloff, A.E. and M. Nord 1993. 'The Forgotten of Northern New England', in *Uneven Development in Rural America,* ed, T.A. Lyson and W.W. Falk, pp. 125–67. Lawrence, KS: University Press of Kansas.

Lynch, O. and K. Talbott 1995. *Balancing Acts: Community-based Forest Management and National Law in Asia and the Pacific.* Washington, D.C.: World Resources Institute.

Mathews, A.S. 2004. 'Forestry Culture: Knowledge, Institutions and Power in Mexican forestry, 1926–2001'. Ph.D. dissertation, Department of Anthropology and School of Forestry and Environmental Studies. New Haven, CT: Yale University.

Moran, E.F. 1981. *Developing the Amazon.* Bloomington: Indiana University Press.

Mosse, D. 1997. 'The Symbolic Making of a Common Property Resource: History, Ecology, and Locality in a Tank-irrigated Landscape in South India', *Development and Change* 28: 467–504.

Mundy, B.E. 1996. *The Mapping of New Spain: Indigenous Cartography and the Maps of the Relaciones Geográphicas.* Chicago, IL: University of Chicago Press.

Newsweek. 30 August 1993.

Northern Forest Alliance 1998. 'Explore the Northern Forest!', (poster) Montpeliet, VT: The Northern Forest Alliance.

Otto, J.S. and N.E. Anderson 1982. 'Slash-and-burn Cultivation in the Highlands South: a Problem in Comparative Agricultural History', *Comparative Study of Society and History* 24: 131–47.

Peluso, N.L. 1995. 'Whose Woods Are These? Counter-mapping Forest Territories in Kalimantan, Indonesia', *Antipode* 27: 383–406.

Pérez García, R. 1996 (1956). *La Sierra Juárez.* Volume 1. Oaxaca, México: Instituto Oaxaqueño De Las Culturas.

Pichón, E.J. 1996. 'Settler Agriculture and the Dynamics of Resource Allocation in Frontier Environments', *Human Ecology* 24(3): 341–71.

Rademacher, A. 1998. 'Restoration as Development: Urban Growth, River Restoration, and Riparian Settlements in the Upper Bagmati Basin, *Kathmandu, Nepal*'. Master's thesis, Yale School of Forestry and Environmental Studies. New Haven, CT: Yale University.

Rangan, H. 1992. 'Romancing the Environment: Popular Environmental Action in the Garhwal Himalayas', in *In Defense of Livelihoods: Comparative Studies in Environmental Action,* ed, J. Friedmann and H. Rangan, pp. 155–81. West Hartford: Kumarian Press.

Reidel, C. 1994. 'The Political Process of the Northern Forest Lands Study', in *The Future of the Northern Forest,* (eds), C. McG. Klyza and S.C. Trombulak, pp. 93–111. Hanover, NH: University Press of New England.

Riles, A. 2004. 'The Empty Place: Legal Formalities and the Cultural State', in *The Place of Law,* (eds), A. Sarat, L. Douglas and M.M. Umphrey, pp. 43–73. Ann Arbor: University of Michigan Press.

Ruthenberg, H. 1976. *Farming Systems in the Tropics.* 2nd edition. Oxford: Oxford University Press.

Schmink, M. and C.H. Wood 1992. *Contested Frontiers in Amazonia.* New York: Columbia University Press.

Schwartzberg, J.E. 1994. 'Introduction to Southeast Asia Cartography', in *The History of Cartography*, Vol. II, Book II: *Cartography in the Traditional East and Southeast Asian Societies*, (eds), J.B. Harley and D. Woodward, pp. 689–700. Chicago, IL: University of Chicago Press.

Scott, J.C. 1998. *Seeing Like a State: How Certain Schemes to Improve the Human Condition Have Failed*. New Haven, CT: Yale University Press.

Sigaut, F. 1979. 'Swidden Cultivation in Europe: a Question for Tropical Anthropologists', *Social Science Information* 18(4/5): 679–94.

Sillitoe, P. 1996. *A Place against Time: Land and Environment in the Papua New Guinea Highlands*. London: Routledge.

———— 2002. 'Globalizing Indigenous Knowledge', in *Participating in Development: Approaches to Indigenous Knowledge*, (ed.) P. Sillitoe, A. Bicker and J. Pottier. London: Routledge.

Simonian, L. 1995. *Defending the Land of the Jaguar*. Austin: University of Texas Press.

Sirait, M., S. Prasodjo, N. Podger, A. Flavelle and J. Fox 1994. 'Mapping Customary Land in East Kalimantan, Indonesia: A Tool for Forest Management', *Ambio* 23(7): 411–17.

Smith, D.S. 2003. 'The Discipline of Nature: a History of Environmental Discourse in the Northern Forest of New England and New York', Ph.D. dissertation, Yale School of Forestry and Environmental Studies. New Haven, CT: Yale University.

Stearman, A.M. 1994. '"Only Slaves Climb Trees": Revisiting the Myth of the Ecologically Noble Savage in Amazonia', *Human Nature* 5(4): 339–57.

Tanaka, M. 1997. *Conditions of Low Income Settlements in Kathmandu: Action Research in Squatter Settlements*. Kathmandu: Lumanti.

Thongchai W. 1994. *Siam Mapped: a History of the Geo-Body of Siam*. Honolulu: University of Hawaii Press.

Thrupp, L.A., S. Hecht and J. Browder 1997. *The Diversity and Dynamics of Shifting Cultivation: Myths, Realities, and Policy Implications*. Washington, D.C.: World Resources Institute.

Tsing, A. 1999. 'Becoming a Tribal Elder and Other Green Development Fantasies', in *Transforming the Indonesian Uplands*, (ed.) T.M. Li, pp. 159–202. Amsterdam: Harwood.

———— 2000. 'The Global Situation', *Cultural Anthropology* 15(3): 327–60.

———— 2005 *Friction: an Ethnography of Global Connection*. Princeton, NJ: Princeton University Press.

Tyrtania, L. 1992. *Yagavila un Ensayo en Ecologia Cultural*. Mexico: D.F. Universidad Autonoma Metropolitana, Unidad Iztapalapa, Division de Ciencias Sociales y Humanidades.

Urry, J. 1990. *The Tourist Gaze: Travel and Leisure in Contemporary Societies*. London: Sage.

Vayda, A.P. 1990. 'Actions, Variations, and Change: the Emerging Anti-essen-
   tialist View in Anthropology', *Canberra Anthropology* 13(2): 29–45.
Wall Street Journal. 'Long Road Back.' 24 November 1989.
Wilmsen, E.N. 1989. *Land Filled with Flies.* Chicago, IL: University of Chicago
   Press.
Wolf, E. 1986. 'The Vicissitudes of the Closed Corporate Community',
   American Ethnologist 13: 325–29.
World Resources Institute. 1996. *The Urban Environment.* New York: Oxford
   University Press.
Zerner, C. 1993. 'Through a Green Lens: the Construction of Customary
   Environmental Law and Community in Indonesia's Maluku Islands', *Law
   and Society Review* 28(5): 1079–1122.
——— (ed.) 2003. *Culture and the Question of Rights: Forests, Coasts, and Seas
   in Southeast Asia.* Durham, NC: Duke University Press.

8 Science and Local Knowledge in Sri Lanka: Extension, Rubber and Farming

Mariella Marzano

Efforts to incorporate local knowledge into natural resource management are growing in Sri Lanka, following a shift in emphasis from 'top-down' technology transfer to more collaborative approaches. Although attractive in principle, these efforts are often fraught with difficulties. Here, I consider the relationship between Western-based 'global' science and local or traditional knowledge in Sri Lanka as mediated through farmer extension services. The ideology and practices of 'outsiders' have largely governed development in Sri Lanka, from British colonial welfare policies to a succession of 'poverty alleviation' programmes instigated by international donor agencies. Sri Lanka has a long relationship with global science in many arenas (e.g. medicine, education) and has an impressive record of agricultural research and support (Pain 1986; Somaratne 2003). Indeed, Research and Development (R&D) in farming feeds into the Sri Lankan government's primary objectives – poverty alleviation, economic growth and the sustainable use of natural resources – through efforts to improve technology and agricultural productivity (Somaratne 2003). While living and working with Sinhalese farmers[1] I contributed to a natural resources R&D project on intercropping with rubber[2] (hereafter called 'the rubber project'). During this work it became clear that a greater understanding of the factors influencing farmer decision making was necessary to increase their uptake of technology and ensure the suitability and sustainability of R&D.

This chapter draws on fieldwork in Moneragala, southeast Sri Lanka. It examines the problems confronted by extension services that, charged with the dissemination of technological advice, operate at the interface between scientific research and local farming practices. The scientists leading the rubber project were transferring their experiments from research stations into local settings in an attempt to encourage greater involvement from local farmers and to discover whether previously successful on-station intercropping experiments with banana would work in the field under differing agro-climatic and socioeconomic conditions. Like many other research facilities in Sri Lanka and worldwide, the Rubber Research Institute faces a demand for technology that is more meaningful and relevant to national needs. Sri Lanka has seen a growing interest in local knowledge and the role that local people can play in rural development (see SLARCIK Report 1996), which is in keeping with the popular notion, that greater participation by local people in their own development, could be the key to a sustainable and more equitable future (e.g., Chambers 1997). Thus, in order to encourage people to grow rubber, projects have to consider diverse socioeconomic and agronomic factors, including the problems of income loss before trees mature and fluctuating prices for latex.

In Sri Lanka, local knowledge and practices are constantly challenging the relevance of global agricultural science (or its application through technology), but what informs the choices farmers make (Uphoff et al. 1990)? It is often difficult to identify how and why villagers accept or reject scientific advice (on an individual or group basis), but it is important to recognise the disparity between the goal of long-term sustainability and the choices that farmers make when adopting local agricultural strategies, usually in the short term (see Arce and Fisher, this volume).

In trying to understand what might prevent the successful participation of local people in natural resource management projects I examine the multiple contexts that influence the decisions people make when choosing to accept or reject global science in the form of development-related advice or technology. Reflecting on the subsequent implications this may have for sustainable development reveals a number of interwoven themes. (1) Sri Lanka's historical relationship with Western knowledge and influence as exemplified by the introduction of rubber and the development of research and extension; (2) the problems of conflicting expectations and the frustrations this causes as Extension Officers are caught between global scientific research and local realities; (3) the role of the wider sociopolitical environment in determining how local people identify their own needs; and (4) the multilayered relationships involved in development, often leading to conflicting extension advice and constraints on local innovation and experimentation.

Rubber in Sri Lanka: Western Influence

Rubber has a long history in Sri Lanka, exemplifying the ways in which the West has influenced and changed ideas and agricultural practices there. Introduced by British colonisers in 1877 (De Silva 1981), its commercial potential was enthusiastically adopted by British planters and entrepreneurs from the Sri Lankan élite over the next twenty years. Indeed, the Administration introduced new modes of economic behaviour, encouraged immigrant workers (largely Indian), and appropriated large tracts of land in the name of the Crown to establish rubber plantations (Alailima 1997; Hochegger 1998).

Introducing rubber as an export crop was one of many Western attempts to control Sri Lanka's natural potential. In constructing a 'superior' knowledge, the British were also imposing control (Rogers 1990; Martin 2003). Senanayake (1984) believes that commercial plantation agriculture influenced the rate at which traditional agricultural practices were modernised. The British discouraged traditional subsistence cultivation such as *chena* (Sri Lankan term for 'shifting cultivation'), believing it to be agriculturally wasteful and because much of the forest land belonged to the Crown under their rule (see also Dove et al. this volume). The British were also responsible for the development of agricultural research in Sri Lanka, establishing the Department of Agriculture in 1912. A range of research stations followed, along with the development of agricultural extension services originally serving the plantation sector, other crops such as rice being added later (Mahaliyanaarachchi 2002; Somaratne 2003).

Moneragala and Its Potential for Rubber

In the district of Moneragala, the second largest in Sri Lanka, the cultivation of rubber on smallholdings is relatively new and farmers, with legal access to at least 0.5 acres (a land permit is necessary to receive a subsidy), are keen to grow it for long-term investment.[3] On maturation, rubber can provide a steady income but intensive intercropping with banana can increase revenue in the five to six years it takes for rubber to become commercially viable. Rubber trees have a lifespan of around thirty years and when senile can be utilised for fuel wood or manufacturing timber. However, despite often low worldwide market prices (reducing the rate of new/re-planting in Sri Lanka) many farmers still choose to grow rubber (Rodrigo et al. 2001).

Moneragala was previously lightly populated due to its relative isolation and problems with malaria. The district is often thought of as a 'punishment post' for government workers because of its isolation and harsh terrain. However, despite the history of poor public investment in the region, land availability attracted farmers who adapted their practices to local natural resource pressures. The population of Moneragala has increased by 60 percent since 1981 largely because of the availability of land (Wanigasundara 1995) but land has since become scarce. Moneragala is considered to be one of the poorest and least developed regions of Sri Lanka and has attracted numerous development-oriented organisations and stakeholders and substantial funding from foreign aid agencies (e.g. UNDP and NORAD).[4] The livelihoods of around 90 percent of the population are connected in some way to farming. However, only 1 percent of school leavers want to farm because of the risk associated with rain-fed cultivation and no guarantee of year-round income (ibid).

The villagers in Therrapahuwa division[5] of Moneragala are mainly settlers from surrounding areas who previously used the fertile jungle for *chena* (shifting cultivation). Except for small plots periodically cultivated on Therrapahuwa Mountain, *chena* has been replaced by non-shifting rain-fed cultivation as land has become scarce. Paddy cultivation is largely rain-fed although some households cultivate 0.5–2 acre plots within nearby irrigated colonies. Smallholders[6] cultivate perennials such as *kesel* (banana), annual and subsistence crops such as *batu* (brinjal, Sri Lankan for eggplant), *mirris* (chillies), *iringu* (maize) and *kurakkan* (millet), as well as long-term plantation-style crops such as teak and rubber. The rubber project's goal was to assist those smallholder farmers who wished to plant, or replant, rubber by suggesting they increase levels of intercropping with banana as a means of both improving land-use efficiency and maintaining the rubber land (see Rodrigo and Stirling 1997). Once this local research was completed, advice would also be disseminated widely to farmers elsewhere via Rubber Development Officers.

The Role of Extension

Extension services in Sri Lanka, which generally operate on a crop-specific basis, play an important role by educating farmers in the use of new agricultural technologies. In practice, Extension Officers are the link between global and local science, mediating between them. They aim to improve agricultural productivity and thereby increase the income level of rural households (Nakamura et al. 1997). However, modern agricultural systems advocated by Western and national administrations sometimes have little relation to the everyday reality of farmer's lives (Moles and Riker 1984). Pain (1986: 755) highlights the fact that agricultural research in Sri Lanka 'offers evidence of a vital and competent research organisation and programme' but others argue that agricultural research has largely failed Sri Lankan farmers:

> The fault lies everywhere. It is ingrained in a system that does not encourage University research to filter down to farmers. In a system that does not protect the farmer adequately from price shocks … there is no mechanism that effectively links the majority of farmers with the cocooned research scientists – in Universities or even in the Agriculture Department's own research stations. (Dissanaike 2000: 7)

The difficulties inherent in making connections between global and local science are compounded by the wider development context. Regardless of much valuable work, rural development efforts in Sri Lanka continually struggle to overcome a long history of paternalism and partisanship which inevitably impacts on sustainable intervention. Development, like many other aspects of Sri Lankan life, is highly political and dominated by a 'top-down' technology-transfer ethic (see Sillitoe, this volume). While there has been a move towards a more positive outlook on the validity of farmer perspectives (Ellen and Harris 2000), it may come as no surprise that connections between scientific research and extension are weak and fragmented. Budgetary constraints,[7] frequent institutional restructuring of R&D,[8] the unequal distribution and quality of Extension Services and a focus on supply- rather than demand-driven programmes can limit the effectiveness of intervention (Somaratne 2003).

Integrating agricultural research and extension is seen as vital for the diffusion of new technologies to local farmers (Somaratne 2003) but the success of extension is limited not only by factors at the macro level (outlined above) but also by the difficulties of working at the local level. The problems of Extension Officers are broadly twofold. First, although they are the contact point for the farmer, they may not have sufficient knowledge of the new technology they are supposed to promote ('top-down' approaches to extension lead to difficulties in disseminating information and providing training for extension workers). They may find it difficult to access villages because of poor transport and rugged terrain or because they are simply too busy with other activities taken on to supplement their low wage (Nakamura et al., 1997). Second, farmers will adopt technology only if the innovation is likely to be of benefit to them. Thus the relationship between Extension Officers and farmers, and the methods the former use to disseminate agricultural information, are important factors in technology uptake. One of the challenges for science and extension is to find a balance between optimal economic productivity and the understanding that smallholder farmers have to meet household needs with limited resources. In Moneragala, the rubber project focused on farmer decision-making to understand why certain crops were chosen to be intercropped with rubber. Research highlighted the challenges facing Extension Officers and their frustrations with farmers who did not follow their advice.

Frustrations of a Rubber Development Officer

One of the tasks of the Rubber Development Officer is to provide farmers with recommendations and advice on intercropping. Intercrops (e.g. banana, brinjal) can be grown with rubber until the fourth year but this practice was not always favoured. Regulations previously enforced by Development Officers (controlling subsidy payments) ensured that monocropping prevailed. Only relatively recently have recommendations been relaxed and intercropping supported as a source of extra income. A rubber subsidy (covering planting material, labour, fertiliser etc.) is an added incentive for farmers to grow rubber, although subsidies are closely linked to extension recommendations (Rodrigo et al. 2001). A field diary extract following a day spent with the Rubber Development Officer highlights the feelings of frustration arising when recommendations confront reality:

> Visited a few of the farmers who were growing rubber. For one acre of rubber land, Sudubanda was due the 8th and last payment. For this the tree had to be 20 inches in girth. Sudubanda, on the advice of a rubber mill owner in the village, start-

ed tapping without informing the Development Officer or waiting till the trees were 20 inches in girth (although Sudubanda insists the trees are 20 inches). By starting to tap before the trees are ready Dharmapala [the Development Officer] said that the undergrowth will stop and the *kiri* (milk or latex) will dry up in 2 years. Luckily Sudubanda stopped tapping as the *kiri* was not enough. The trees Sudubanda had started to tap, when measured by Dharmapala, were only 14.5 inches. Sudubanda said that 120 [out of 200] of his trees were of tappable girth but Dharmapala estimated that only 15–20 were … they can only approve the last payment if 60% is of tappable girth. He thinks that Sudubanda started to tap early to try and earn some money. Sudubanda will probably not get his 8th subsidy payment as the girth is not tappable until next year and there is a 6 year deadline to get the trees to tappable girth. [We visited another farmer]… he had cleared between the trees for brinjal cultivation but not around them nor cut the side branches. Dharmapala said that Jayasingha was concentrating on brinjal instead of rubber because he can get an income from brinjal. Thus he hasn't maintained the rubber. Gunapala fared a little better although apparently he hadn't taken Dharmapala's advice and dug drains for soil conservation. However he had intercropped brinjal with rubber and kept the land cleared. Dharmapala said that intercropping with a cover crop is necessary. This is the message he relays to farmers. He is not supposed to give subsidies if trees are not the recommended girth but realises that the farmers sometimes have extenuating circumstances. If girth is not good he gives advice to use more fertiliser or weed the land. However, Dharmapala is not happy. He came over in the evening … He said that there was a lot of development money in the region but no development … this was the fault of Government servants and farmers who were only interested in money and not developing their lands.

Dharmapala was frustrated with villagers who had not fully followed his recommendations, pointing out that he also faced difficulties of competing priorities but was still able to cultivate his land despite a full-time job. One of the main problems, he thought, was that he had to cover a large geographical area and was not able to visit farmers to offer advice as often as he would like.

The day spent with Dharmapala highlighted some of the confusing aspects of development at the grassroots. Many villagers wanted to grow rubber (and receive often-delayed subsidies) but most did not maintain the land or rubber as recommended by Development Officers. They gave a variety of reasons for not following the recommendations, including illness, childbirth and other harvesting priorities. However, this exchange also highlights the interesting question of what governs which advice people take and which they reject? To come closer to an understanding of how farmers make decisions, the frustrations of the Rubber Development Officer must be set within a wider context of development agendas in Sri Lanka and the ways in which these agendas are played out in the local setting. Before returning specifically to rubber, I now want to offer four wider perspectives involving extension, each affecting the decisions that farmers make.

Development Agendas and Local Realities

While current agricultural development generally focuses on sustainable solutions to environmental and social problems, the extent to which villagers are willing and able to accept advice varies. Conflicting expectations of farmers, scientists and policy makers (with the supporting workers in between) over defining 'appropriate' development remains problematic in Sri Lanka (see also Sillitoe 1998a). Another field diary extract reveals the local reality of some development efforts for farmers.

> Called for Nireka at 8.45 am and we walked down to the school as people from one of the University's Agricultural Department were coming to talk about cultivation. We waited at Sudu Menike's house until the agricultural people turned up. They ... are touring the country ... Around 22 farmers (roughly 15 men and 7 women) were there and the hall was also filled with children. Apparently the agricultural department was going to hand out some seeds and shoots. Whilst waiting around outside for them to set up one of the women said she wanted chilli seeds. She said a lorry was going to turn up later with some produce but meanwhile videos would be played on agricultural practices. The first video was on paddy and lasted for three quarters of an hour ... The school children were making a bit of noise and by the second video most of the male farmers had left. Nireka said that it was because farming practices were different here, there isn't enough land for everyone to have paddy and farmers only wanted shoots and seeds anyway ... No shoots or seeds were given but there were books [instruction manuals] for sale.

Villagers' expectations were different from those of the visitors ('outsiders'[9]) who tried encouraging farmers to increase paddy yields using a mixture of traditional practices (natural fertilisers, home-made devices for dealing with pests) and regulated chemical fertilisers and pesticides. One of the Officers explained why the mobile unit had come: it was one way of quickly disseminating agricultural technology and ideas. While this visit was part of an 'awareness programme', the farmers clearly did not consider the extension advice to be relevant for the region and thus were not motivated to listen. Interestingly, the video used by the visiting Extension Officers plugged into nationalist ideologies, stressing that 'During the King's period Sri Lanka was self sufficient in rice ... even export[ing] rice. We had a strong economy which was built up on the tank and dāgäba culture'.[10] The video further emphasised that the unity of villagers would contribute to agricultural and economic success. As previous anthropological studies of rural Sri Lanka show, rural development discourses are closely linked to the dissemination of nationalist ideologies, incorporating notions of village unity and harmony and adherence to Sinhalese Buddhist doctrine (Brow 1988; 1990; Moore 1985, 1992a; Spencer 1990; Woost 1990, 1993). This nationalist discourse celebrates the long history of peasant farmers using a highly selective construction of a glorious national past (under Sinhalese kings) to promote a model of future development (Brow 1988). Rural villagers become ever more incorporated into larger political and economic systems as agriculture is increasingly influenced by bureaucrats (Brow 1988; Upawansa and Wagachchi 1999).

The danger here is that notions of development are being channelled through extension to impose politically motivated agendas not necessarily responsive to people's needs or their understanding of their livelihood problems (Grillo and Stirrat 1997; Mosse 2001). Moreover, the message that development can be achieved through unity and harmony is in direct contrast to the way villagers have been conditioned to view intervention. As people struggle over access to limited resources (Brow and Weeramunda 1992), villagers in Moneragala have become highly politicised and are adept at using, manipulating and negotiating their interactions with others. A long history of political wrangling, partisanship and corruption has also had a profound impact on sustainable development interventions that seek participation from local people, a process that requires a climate of trust and cooperation (see Sillitoe, chaptaer 1, this volume). The devastating impact of civil war and past JVP[11] insurgencies on the lives of Sri Lankans must also be remembered: violence has influenced how people view themselves and others (Spencer 1990; Moore 1992b; Alailima 1997). In reality, there exists a widespread inability to foster collective action for development as communities are divided by caste, class and party allegiance (Gunasekara 1992). Supporting the right party can lead to lucrative projects and jobs, and, contrary to the participatory ideology of current development discourse that aims to tackle inequality, rural villagers often believe that access to resources comes from connections with outsiders (Woost 1997).

Shifting Concepts of Local Knowledge and Sustainability

Traditional Sri Lankan agriculture is accompanied by a wide range of spiritual, supernatural and astrological beliefs and rituals (Handawela 2001) and it is crucial to understand that farmers' local knowledge comes from many sources both historic and current. Sri Lankans operate in a social environment long open to the flow of people, goods and knowledge (Argenti-Pillen 2003) and, within the villages, farming knowledge involves a mixture of individual experimentation, advice handed down through the generations or exchanged between neighbours and kin. Global science and technology also reaches local farmers through various sources including extension (see also Dove et al., Arce and Fisher, Sillitoe, this volume).

Some (e.g., Upawansa and Wagachchi 1999) believe that indigenous farming systems are disappearing and that the introduction of science-based education has accelerated this process. However, local knowledge is not static but constantly evolving and so incorporates an understanding and manipulation of broader sociopolitical factors as well as recommendations from Extension Officers (Sillitoe 2000a). This is particularly relevant in the context of development. In the recent past, farmers have been encouraged to adopt environmentally unsustainable practices through the use of chemical fertilisers and pesticides because these translate into higher crop yields and income in the short term. However, such development strategies influenced by global science have long been criticised for ignoring local knowledge and instituting unsustainable practices (Sundar 2003). Now such practices have fallen from favour and farmers are requested to conform to the latest ideas on environmentally sustainable agriculture. This results in a disparity between the assumptions of development regarding sustainability and farmers' previous experiences of managing market uncertainty.

Farmers are well aware of the development discourse, which labels those that do not adopt current 'good practice' (in the form of extension recommendations) as 'lazy', 'ignorant' or 'backward' and use this language themselves (see also Woost 1993). On the other hand, it is extremely difficult for many of them to break out of the cycle of agri-chemical use. For example, the favoured cash crop in Therrapahuwa is brinjal. It is preferred over banana as an intercrop with rubber, and I took a particular interest in it. Farmers, were well aware of the environmental problems caused by their use of agri-chemicals but did not have the time nor resources to make enough natural compost to cover acres of brinjal. Moreover, the large fluctuations in brinjal prices would dissuade anyone from such an effort. As this diary extract reveals, local farmers were keen to point out that if they were *given* compost or saw a quick economic benefit they would use it:

> Chanti said there had been a meeting at Dingiribanda's (while I was in Colombo) on *batu* (brinjal) cultivation. Apparently cultivating brinjal year after year has meant that the soil is not productive and yield is low. They were promoting the use of compost as fertiliser. Chandra said that everyone knows how to do it and thought it was a good idea but nobody had done it. Chandra said you are tired making compost and then you get nothing for brinjal (she picked 70 kilos today – middleman price 6 rupees a kilo). She added that farmers are lazy and they want the easy option but it would be an incentive to try natural fertilisers if the price was good.

Sri Lankan 'outsiders' such as Extension Officers face particular problems with paternalistic expectations in attempting to improve local farmers' lives by encouraging them to adapt more sustainable practices. Officers are now trying to expound current international 'environmentally friendly' practices to an audience that mostly views potential development in a paternal or patron/client fashion (Brow 1990; Moore 1992b; Woost 1993; Wickramasinghe 2001) and are accustomed to using agri-chemicals. A culture of 'quick rewards' is intensified further by some extension recommendations (often both causing and influenced by short-term trends), which can also lead to conflicting advice and unsustainable practices.

Short-term Trends

The 'unsustainable' environmental practices adopted by farmers have been exacerbated by the short-term trends of extension advice and market fluctuations. For example, there is a constant flooding of markets with the crops promoted by extension. Like many Sri Lankan villages, those I know have seen their fair share of development projects, most of which had fallen by the wayside. Moneragala shows a pattern, apparently repeated across Sri Lanka, where a specific crop (e.g. pepper, maize) is endlessly promoted during a particular year by various government departments or organisations. Farmers are encouraged to adopt the relevant technology by promises of high prices, offers of loans or subsidies. Many accept and those who do so early generally get some benefits though, inevitably, most fail to make any money as markets soon flood and prices drop. Farmers then switch to the next 'crop of the year', stop cultivating the previous one and leave behind unpaid loans. This cycle inevitably affects the mindset of Sri Lankans, who are now resigned to short-term projects which may yield some quick rewards but which, in the long run, are unsustainable (see also Gegeo 1998). The following diary extract illustrates why many villagers are selective in their response to rural development and extension efforts:

> Mr Gunasena (retired villager elder) complained about the Government. He said that everything is done without a system in Ceylon. He told of going to a funeral house where he met the GA (Government Agent – a friend of his). This GA admitted that the price of something is high, it is grown, it is attacked by disease and then the price becomes low and it is abandoned. Mr Gunasena said that when he came here there was a *belun* [a type of onion] scarcity. The agricultural Extension Officer advised them to cultivate this. In a year it was over. The next year it was red onion but the price became too low, the next year it was soya so all the farmers cultivated soya, then sorgum … At its introduction the advantages of cultivating it were so high and then after a year it was over … He believed the Government and the middle men to be conspiring together. Mr Gunasena and Dharamadarsa [who later joined the conversation] said there was no protective system. The farmer is at the mercy of the middle man. Mr Gunasena said the quality of the produce may be the best but the price still belongs to the middle man.

Seeds of Confusion: Conflicting Advice

In recent years, agricultural Research and Development in Sri Lanka has increasingly involved private sector extension services in the form of NGOs or commercially-related enterprises, in response to the failure of the state to run an effective, integrated extension service (Somaratne 2003). This is not to say that there is no state provision for villagers. Smallholder cultivation in Sri Lanka is supported by a wide range of government services (Moore 1985) but government agencies in Moneragala face difficulties in effectively delivering these services to farmers. For example, high staff turnover sometimes affects the 'institutional memory' of agencies, leading to mistakes being repeated. Agencies are also influenced by national problems. Sri Lanka's public administration is characterised by devolution of power to regional authorities (Marzano 2002) which, instead of increasing communication between government departments and other development-related organisations, leads to considerable uncertainty (see Moore et al. 1995; Lakshman 1997). Bond (2003: 90) highlights the conundrum facing development in Moneragala:

> What was lacking was any institutionalised ability to learn from either success or failure of attempted solutions; and any co-ordinated planning system. There was also a marked compartmentalisation of effort, agencies neither communicated nor co-operated. There were deep attitudinal obstacles, which posed a threat to any project hoping to achieve these ambitious ends within the short to medium term ... The state had a strong ethos of control, regulation and provision for the people within a bureaucratic culture where risk taking is discouraged ... The NGOs were numerous but isolationist and arrogant, considering that they alone and individually were capable of improving the lot of the poor, rejecting the perceived 'error' of other NGOs, 'corruption' of government and 'exploitation' of business.[12]

While on the surface a wider range of development-related organisations may suggest that more farmers can be reached, the diversity of 'science-based' extension advice (each with its own motivations and agendas) can disrupt the adoption of ideas by farmers. The resulting confusion amongst farmers on the receiving end of often-conflicting advice militates against sustainable development. For example, a government agent recounts a village incident to me where three bodies – an NGO, a government department and a tobacco company – fought over the right to 'develop' the same farmers:

> The [tobacco company] encourages farmers to use powerful chemicals, pesticides and fungicides. Also conducting development with the same farmers is an NGO ... who encourages farmers to use bio-methods. Also conducting development with some of the farmers is the [Government] Department – encouraging farmers to use correct guidelines – for example – correct amount of fertilisers, pesticides etc. So with the same target farmers you have three groups encouraging them to do different things. One saying 'use biodiversity, organic methods!', the second says, 'use regulation chemicals!', the tobacco company is saying 'use as many chemicals as possible for lovely large leaves!' Who are farmers going to choose? What is the most sustainable?

This example should not detract from the valuable work of individuals within State departments and NGOs in Moneragala but it does show that all are part of a 'development hierarchy' that has links to the global arena (Woost 1997). It is these exchanges that influence the perceptions that farmers have of development and, subsequently, the choices that they make.

A Return to Rubber Intercropping

Awareness of the sociopolitical environment in which global and local science interact is clearly important for understanding why some farmers adopt some extension recommendations and reject others as well as the extent to which they are able to apply their own local knowledge and experiment and innovate for themselves. Although some believe that modern science and agricultural scientists have little faith in local knowledge (e.g. Upawansa and Wagachchi 1999), this is certainly not always so. The rubber project considered optimising the planting density of an appropriate intercrop between rows of rubber to be a practical, low-input and adaptable technology precisely because it presumed farmers already had extensive knowledge of native plants. One villager who had just planted rubber, when asked what he needed from 'outsiders', said:

> If the Government gives something like rubber we want information ... Officers come to see and give advice on how to use fertiliser, how to plant, medicine for plants. We don't want advice about intercrops ... we know about vegetables.

The rubber project took a strong interest in the vegetables that people chose for intercropping. This interest revealed two further issues relevant to a discussion of global and local science: the tacit nature of farmers' knowledge and the provision, and resulting constraints, of subsidies.

The location chosen by farmers for intercropping plants was of particular interest to the project's scientists as an aspect of local decision making but discussions on this topic were often difficult (see also Sillitoe 1998b). When asked why they had planted certain species in specific locations the villagers would give an explanation that generally related to their particular home garden but often answered *nikam* ('just because'). Clearly such tacit knowledge is not something a farmer can easily communicate (see Heckler, this volume). Furthermore, while espousing the necessity to integrate local knowledge in rural development efforts, policy makers are often unable to work with (or use) 'complicated results' that arise when working from the farmer's perspective (Cain *et al.* 2003). Local science may thus effectively become inaccessible and the associated danger here is that it is ignored or misunderstood (Sillitoe 1998b). This, in turn, may contribute to the belief that indigenous farming systems are disappearing as a result of globalised scientific education. On the contrary, local science should be considered as being a 'dynamic hybridisation with the wider world', not static but constantly adapting (Martin 2003: 59 and references therein, Dove et al. this volume).

Whatever intercrop a farmer chooses to plant during the immature stages of rubber depends on such factors as the intercrop's market value and the need for land to grow traditional cash crops and food for domestic use. When the rubber is mature, a farmer's decision to allow it to stand alone or to intercrop with cocoa again depends to a large extent on the availability of labour and the possibility of receiving subsidies for the intercrop. Decisions over intercropping with mature rubber are among the most visible sources of experimentation by farmers in Moneragala. On the other hand, the rules governing extension work can discourage or prevent farmers from experimenting and innovating for themselves. This arises specifically from the imposition of conditions associated with the provision of subsidies that, ultimately, constrain farmers to following extension recommendations.

The Rubber Development Department encourages farmers to cultivate cocoa as an intercrop as there is a chance they will receive subsidies.[13] Despite this encouragement, farmers in Moneragala whose rubber trees are reaching maturity do not want to take up this option because squirrels commonly attack the cocoa plants. Although squirrels usually only eat the flesh from around the cocoa seed, farmers generally discard those seeds that fall. As an alternative, rubber-growing smallholders have also experimented by planting shade-tolerant plants (e.g. pepper and coffee) despite being heavily discouraged by the Development Officers, largely because research on the suitability of these plants was not complete and thus these potential intercrops were not yet approved.

Through controlled experiments, science is reasonably confident of cause and effect but when experimentally derived scientific 'facts' are transferred to the field they inevitably face wider interpretation and both environmental and human variation. Richards (1993) notes that agricultural researchers are powerful individuals whose confidence (that small-scale farmers would perform better if they took scientific advice) may discourage farmers from voicing their own opinions or scepticism. Nonetheless, Novellino (2003: 273) has demonstrated that local people may devise their own strategies as a defence against the dominant discourse. He highlights that 'miscommunication' between local people and those who represent development is common and stems from 'attitudes, stereotypes and stigmatisation'. However, this should not be viewed as failed communication, rather as a form of negotiation. On the other hand, farmers' defensive strategies may be ineffective in relation to their freedom to make choices when global science, bolstered and directed by the provision of subsidies offered through extension services, severely affects the application of local knowledge by discouraging farmers from experimentation and innovation.

Extension, Science and Development

By drawing on examples from fieldwork in Sri Lanka and wider literature, I have tried to highlight the many factors that influence effective collaboration at the interface between global and local science. The relationship between Sri Lanka and the West is set within a climate where continuing patronage is coupled with a new drive for investment-led development. This is located against a highly charged political situation where nationalist (Buddhist) ideologies of a traditional Sinhalese past spill over into the discourse of many rural development programmes (Woost 1993). As Sillitoe (2000b: 9) points out, 'a tendency to see indigenous knowledge as saving cultural property from being lost may be related to deep-rooted identity issues'. Sri Lanka's colonial history and enduring relationship with the West continues to influence the ways in which development is formulated and implemented. Yet the perpetuation of British colonial attitudes post-independence, coupled with romanticised paternal outlooks on rural life, has created growing tensions given the current hyper-capitalism in Sri Lanka (Winslow 2002). Farming can thus become an arena for playing out political tensions. Indeed, many villagers have become highly politicised and, aware of development agendas, are decidedly selective in their responses to intervention.

The current favour of participatory approaches and the inclusion of local knowledge in agricultural development have been steadily growing in Sri Lanka. However, moves within agricultural R&D towards greater interaction with local farmers has highlighted problems of conflicting expectations and growing frustrations when science meets local realities. Incompatible and contradictory development agendas serve as a constraint on effective collaboration between global and local science. The popularity of development and the sheer number of agencies vying with each other to 'develop' farmers often leads to institutional confusion. Thus individuals, particularly Extension Officers, face problems working within a system that does not have suitable mechanisms to allow them to respond to specific local, rather than generalised, needs. Furthermore, agencies are not always clear about how to achieve their extension goals, a predicament that might be exacerbated by the need both to justify themselves to the wider world (politically, philosophically and economically) and to explain themselves in practical, accessible and accountable terms in local contexts. A more collaborative approach, to which the development world attaches notions of long-term sustainability based on local needs and wants, can also be interpreted differently by local people who may only 'participate' to the extent that they are willing and able, to see immediate rewards.

Nader (1996: 3) highlights the fact that the boundaries of science are constantly drawn and redrawn but that 'to demarcate science from the problems it confronts, from the solutions it engenders is to remove it from context'. My aim has been to highlight the complex arena within which knowledges (more often hybrid than not) are understood, misunderstood and negotiated. While any knowledge is inherently culturally located (Agrawal 1995, Sillitoe 2000a, b), it can also be influenced through a multitude of sources from the local to the global. Research carried out for the rubber project provided a forum for highlighting that interactions between actors with different agendas and values can lead to tensions that work against collaborative processes necessary for sustainable development (Sillitoe 2000a). However, successful collaboration and effective extension are perhaps also down to the personalities of those involved. Bond (2003) considers that many of the problems in Moneragala are rooted in individual attitudes and many opportunities lie in individual knowledge. Understanding and working with local farmers is not easy but there is an urgent need for global science, mediated through extension, to understand and accommodate the multi-faceted nature of farmer decision making, experimentation and innovation at the local level. Continuing efforts to tease apart the deeply interwoven strands that determine how local people derive and maintain their livelihoods is important if development intervention is to take on board the many factors that influence the decisions that farmers make.

Notes

1. The Sinhalese make up around 74 percent of the population (Department of Census and statistics www.statistics.gov.lk)
2. This chapter is an output from a project (Plant Science Research Program 7212) funded by the U.K. Department for International Development (DFID)and administered by the Centre for Arid Zone Studies. The views expressed here are not necessarily those of DFID or other colleagues involved in the project.
3. Farmers who do not hold a land permit or enough land on which to cultivate rubber can earn an income by intercropping on the rubber lands of another villager in return for maintaining the trees (Janowski, 1997).
4. UNDP – United Nations Development Project; NORAD – Norwegian Agency for Development Cooperation.
5. The Sri Lankan villages of Mediriya, Therrapahuwa and Walamatiara, where I carried out my research, are part of the Grama Seva division of Therrapahuwa, comprising six, predominantly Buddhist, villages. One of forty-two administrative branches included in the Badulkumbura Divisional Secretariat, this division in Uva province is one of ten largest administrative units within Moneragala.
6. Smallholdings can be an extension of home gardens or be on land in another part of the village, usually belonging to someone else.
7. Heavy government spending on the civil war led to reduced government spending on R&D in agriculture (Somaratne 2003). However, there have also been accusations that government policymakers are uninterested in extension and provide little investment or support (Mahaliyanaarachchi, 2002).
8. For a more comprehensive overview of the evolution of Extension Services see Mahaliyanaarachchi (2002) and Somaratne (2003).
9. 'Outsiders' is a very fluid category but, in the context of extension workers, this term could refer to a person being from another village or region, having an education or simply representing an organisation or a government department.
10. Woost (1993: 504) states that 'several interconnected symbolic themes emerge from this ideological formation: the *väva* (the village irrigation tank used in paddy cultivation); the *yāya* (the paddy field itself) and the *dāgäba* (the 'stupa' or religious structure that metonymically represents Sinhala Buddhism)'.
11. Janatha Vimukthi Peramuna – People's Liberation Front.
12. See Stirrat and Henkel (1997), Wickramasinghe (2001).
13. Cocoa is under the jurisdiction of the Department of Export Agriculture which has its own extension services under the Integrated Agricultural Extension Services IAES (Mahaliyanaarachchi 2002).

References

Agrawal, A. 1995. 'Dismantling the Divide between Indigenous and Scientific Knowledge', *Development and Change* 26: 413–39.

Alailima, P.J. 1997. 'Social Policy in Sri Lanka', in *Dilemmas of Development: Fifty Years of Economic Change in Sri Lanka.* (ed.) W.D. Lakshman, pp. 127–70. Colombo: Sri Lanka Association of Economists.

Argenti-Pillen, A. 2003. 'The Global Flow of Knowledge on War Trauma: the Role of the 'Cinnamon Garden Culture' in Sri Lanka', in *Negotiating Local Knowledge: Power and Identity in Development,* (eds). J. Pottier, A. Bicker and P. Sillitoe, pp. 189–214. London: Pluto Press.

Bond, R. 2003. 'Opening Pandora's Box: Regional Action on a Sustainable Growth with Equity Concept', in *Development Planning and Poverty Reduction,* ed, D. Potts, P. Ryan and A. Toner, pp. 86–93. Houndsmills: Palgrave Macmillan.

Brow, J. 1988. 'In Pursuit of Hegemony: Representations of Authority and Justice in a Sri Lankan Village', *American Ethnologist* 15: 311–27.

——— 1990. 'Nationalist Rhetoric and Local Practice: the Fate of the Village Community in Kukulewa', in *Sri Lanka: History and the Roots of Conflict,* (ed.) J. Spencer, pp. 125–44. London: Routledge.

Brow, J. and J. Weeramunda (eds) 1992. *Agrarian Change in Sri Lanka.* London and New Delhi: Sage Publications.

Cain, J.D., K. Jinapala, I.W. Makin, P.G. Somaratna, B.R. Ariyaratna and L.R. Perera 2003. 'Participatory Decision Support for Agricultural Management: a Case Study from Sri Lanka', *Agricultural Systems* 76: 457–82.

Chambers, R. 1997. *Whose Reality Counts? Putting the First Last.* London: ITDG Publishing.

Department of Census and Statistics – Sri Lanka: http://www.statistics.gov.lk [accessed 1 June 2003].

De Silva, K.M. 1981. *A History of Sri Lanka.* Chennaim: Oxford University Press.

Dissanaike, T. 2000. 'Farmer in the Dark', *The Sunday Times* 13 August, p. 7.

Ellen, R. and H. Harris 2000. 'Introduction', in *Indigenous Environmental Knowledge and Its Transformations: Critical Anthropological Perspectives,* ed, R. Ellen, P. Parkes and A. Bicker, pp. 1–34. The Netherlands: Harwood Academic Publishers.

Gegeo, D.W. 1998. 'Indigenous Knowledge and Empowerment: Rural Development Examined from Within', *The Contemporary Pacific,* 10(2): 289–315.

Grillo, R. and R.L. Stirrat (ed) 1997. *Discourses of Development: Anthropological Perspectives.* New York and Oxford: Berg.

Gunasekara, T. 1992. 'Democracy, Party Competition and Leadership: the Changing Power Structure in a Sinhalese Community', in *Agrarian Change in Sri Lanka*, (eds), J. Brow and J. Weeramunda, pp. 229–60. London and New Delhi: Sage Publications.

Handawela, J. 2001. 'Towards a Methodology to Test Indigenous Knowledge', *Compas Magazine* (March).

Hochegger, K. 1998. *Farming Like the Forest: Traditional Home Garden Systems in Sri Lanka*. Germany: Margraf Verlag.

Janowski, M. 1997. 'Report on an Analysis of Rural Livelihoods and Poverty in Two Villages in Sri Lanka', Kent, Natural Resources Institute.

Lakshman, W.D. 1997. 'Income Distribution and Poverty', in *Dilemmas of Development: Fifty Years of Economic Change in Sri Lanka*, (ed.) W.D. Lakshman, pp.171–222. Colombo, Sri Lanka Association of Economists.

Mahaliyanaarachchi, R.P. 2002. 'Agricultural Extension Service in Sri Lanka', *BeraterInnen News* 2: 10–15.

Martin, A. 2003. 'On Knowing What Trees to Plant: Local and Expert Perspectives in the Western Ghats of Karnataka', *Geoforum* 34: 57–69.

Marzano, M. 2002. 'Rural Livelihoods in Sri Lanka: An Indication of Poverty?' *Journal of International Development* 14: 817–28.

Moles, J.A. and J.V. Riker 1984. 'Hope, Ideas, and Our Only Alternative – Ourselves and Our Values: National Heritage and the Future of Sri Lanka Agriculture', in *Agricultural Sustainability in a Changing World Order*, (ed.) G.K. Douglass, pp. 239–70. Boulder, CO: Westview Press.

Moore, M. 1985. *The State and Peasant Politics in Sri Lanka*. Cambridge: Cambridge University Press.

—— 1992a. 'The Ideology and History of the Sri Lankan "Peasantry"', in *Agrarian Change in Sri Lanka*, (eds), J. Brow and J. Weeramunda, pp. 325–56. London and New Delhi: Sage Publications.

—— 1992b. 'Sri Lanka: a Special Case of Development', in *Agrarian Change in Sri Lanka*, (eds), J. Brow and J. Weeramunda, pp. 17–40. London and New Delhi: Sage Publications.

Moore, M., Y. Rasanayagam and K.W. Tilakaratne 1995. *Moneragala. Integrated Rural Development Programme (Sri Lanka): Mid Term Programme Review.* Report to NORAD (Norwegian Agency for Development Cooperation) and Ministry of Finance, Planning, Ethnic Affairs and National Integration, Government of Sri Lanka.

Mosse, D. 2001. '"People's Knowledge", Participation and Patronage: Operations and Representations in Rural Development', in *Participation: The New Tyranny*, (eds), B.Cooke and U. Kothari, pp. 16–35. London: Zed Books.

Nader, L. 1996. *Naked Science: Anthropological Inquiry into Boundaries, Power and Knowledge*. London: Routledge.

Nakamura, H., P. Rathnayake and S.M.P. Senanayake 1997. 'Agricultural Development: Past Trends and Policies', in *Dilemmas of Development: Fifty Years of Economic Change in Sri Lanka*, (ed.) W.D. Lakshman, pp. 250–92. Colombo: Sri Lanka Association of Economists.

Novellino, D. 2003. 'From Seduction to Miscommunication: The Confession and Presentation of Local Knowledge in "participatory development"', in *Negotiating Local Knowledge: Power and Identity in Development*, (eds), J. Pottier, A. Bicker and P. Sillitoe, pp. 273–97. London: Pluto Press.

Pain, A. 1986. 'Agricultural Research in Sri Lanka: a Historical Account', *Modern Asian Studies* 20(4): 755–78.

Richards, P. 1993. 'Cultivation: Knowledge or Performance?', in *An Anthropological Critique of Development: The Growth of Ignorance*. (ed.) M. Hobart, pp. 61–78. London: Routledge.

Rodrigo, V.H.L. and C.M. Stirling 1997. *Effects of Population Density on Seasonal Light and Water Use of Banana-Based Interculture Systems*. Report for the Institute of Terrestrial Ecology, Bangor.

Rodrigo, V.H.L., S. Thennakoon and C.M. Stirling 2001 'Priorities and Objectives of Smallholder Rubber Growers and the Contribution of Intercropping to Livelihood Strategies: a Case Study from Sri Lanka', *Outlook on Agriculture* 30(4): 261–66.

Rogers, J.D. 1990. 'Historical Images in the British Period', in *Sri Lanka-History and the Roots of Conflict*, (ed.) J. Spencer, pp. 87–106. London: Routledge.

Senanayake, R. 1984. 'The Ecological, Energetic, and Agronomic Systems of Ancient and Modern Sri Lanka', in *Agricultural Sustainability in a Changing World Order*, (ed.) G.K. Douglass, pp. 227–38. Boulder, CO: Westview Press.

Sillitoe, P. 1998a. 'The Development of Indigenous Knowledge: a New Applied Anthropology', *Current Anthropology* 39(2): 223–52.

——— 1998b. 'Knowing the Land: Soil and Land Resource Evaluation and Indigenous Knowledge', *Soil Use and Management* 14: 188–93.

——— 2000a. 'Cultivating Indigenous Knowledge on Bangladesh Soil: an Essay in Definition', in *Indigenous Knowledge Development in Bangladesh: Present and Future*, (ed.) P. Sillitoe, pp. 145–60. London: Intermediate Technology Publications.

——— 2000b. 'The State of Indigenous Knowledge in Bangladesh', in *Indigenous Knowledge Development in Bangladesh: Present and Future*, (ed.) P. Sillitoe, pp. 3–22. London: Intermediate Technology Publications.

SLARCIK (Sri Lanka Resource Centre for Indigenous Knowledge) 1996. 'Indigenous Knowledge and Sustainable Development'. Proceeding of the First National Symposium, held 19–20 March 1994, University of Jayewardenapura, Sri Lanka.

Somaratne, W.G. 2003. 'Country Paper: Sri Lanka', in *Integration of Agricultural Research and Extension*, (ed.) R. Sharma, pp. 238–53. Report of the APO Study Meeting on Integration of Agricultural Research and Extension, Philippines, 18–22 March 2002.

Spencer, J. (ed.) 1990. *Sri Lanka: History and the Roots of Conflict*. London: Routledge.

Stirrat, R.L. and H. Henkel 1997. 'The Development Gift: the Problem of Reciprocity in the NGO World', *Annals AAPSS* 554: 66–80.

Sundar, N. 2003. 'The Construction and Destruction of "Indigenous" Knowledge in India's Joint Forest Management Programme', in *Indigenous Environmental Knowledge and its Transformations: Critical Anthropological Perspectives*. (eds), R. Ellen, P. Parkes and A. Bicker, pp. 79–100. The Netherlands: Harwood Academic Publishers.

Upawansa, G.K. and R. Wagachchi 1999. 'Activating All Powers in Sri Lanka Agriculture', in *Food for Thought: Ancient Visions and New Experiments of Rural People* (Compas), (eds), B. Haverkort and W. Hiemstra, pp. 105–22. London: Zed Books.

Uphoff, N., M.L. Wickramasinghe and C.M. Wijayaratna. 1990. '"Optimum" Participation in Irrigation Management: Issues and Evidence from Sri Lanka', *Human Organization* 49(1): 26–40.

Wanigasundara, M. 1995. *The MONDEP Decade: a Document on the Integrated Rural Development programme, Moneragala*. MONDEP.

Wickramasinghe, N. 2001. *Civil Society in Sri Lanka: New Circles of Power*. New Delhi: Sage Publications.

Winslow, D. 2002. 'Co-opting Cooperation in Sri Lanka', *Human Organization* 61(1): 9–20.

Woost, M.D. 1990. 'Rural Awakenings: Grass Roots Development and the Cultivation of a National Past in Rural Sri Lanka', in *Sri Lanka: History and the Roots of Conflict*, (ed.) J. Spencer, pp.164–83. London: Routledge.

——— 1993. 'Nationalising the Local Past in Sri Lanka: Histories of Nation and Development in a Sinhalese Village', *American Ethnologist* 20(3): 502–21.

——— 1997. 'Alternative Vocabularies of Development? "Community" and "Participation" in Development Discourse in Sri Lanka', in *Discourses of Development: Anthropological Perspectives*, (eds), R.D. Grillo and R.L. Stirrat, pp. 229–53. New York and Oxford: Berg.

9 Creating Natural Knowledge: Agriculture, Science and Experiments

Alberto Arce and Eleanor Fisher

Introduction

Contemporary anthropology has contributed to a process of reflection on the possibilities and limitations of science, helping to identify the significance of non-scientific knowledge – e.g., indigenous, traditional, and local – in people's understandings of the world (e.g., Warren et al. 1995; Richards 1985; Sillitoe 1998; Ellen and Harris 2000). This has served to underline that processes of knowledge creation and negotiation do not simply belong to a scientific domain. To critically reflect on the boundaries of knowledge and to go 'beyond science' suggests that anthropology can play a role in the development of conceptual approaches that capture and expose the contradictory character of scientific knowledge in the construction of modern contemporary lives and global processes.

In this context, the concept of 'natural knowledge' acquires significance. In the seventeenth century, the term natural knowledge was used variously to describe 'a state of knowing', 'the mastery of learning' or 'a skill or craft' (Wright 1981: 83). Later its dominant meaning became 'knowledge about the workings of nature', having been adopted in the full title of the Royal Society.[1] In recent years, anthropologists and historians seeking to deconstruct science have used the term to refer either to scientific data before the late nineteenth century, treating it as synonymous with natural philosophy (Wright 1981: 97; Dear 1990), or to suggest a process of questioning assumptions about the nature of science and its relationship to human culture and to non-human elements[2] in contemporary life.

In this paper the notion of 'natural knowledge' is used to refer to the rebounding effect of rational knowledge and scientific procedures in relation to human creativity and experience. As such, natural knowledge is not a type of knowledge in its own right. By 'rebounding' we draw on Parkin's (1995: 144) idea of the way in which knowledge practices are constructed and transposed or re-accentuated in different contexts irrespective of whether they emanate from global or local contexts (see also Arce and Long 2000: 8). This implies an emphasis on how scientific rationality, rather than being dissociated from human action, is combined with other factors, such as skill, worldview, verbal definitions and shared practices. In this sense, using the concept of natural knowledge enables us to focus on processes in which different knowledges come together and combine with different practices, rather than placing the emphasis on typologies of knowledge.

The historical referents of natural knowledge in the seventeenth century are seen as significant because historically the establishment of natural knowledge was part of a process in which the boundaries between knowledge generated through everyday life and knowledge generated within institutions (such as the Church or the predecessors of scientific establishments: medicine and the universities) were extremely fluid and blurred, permitting debate about where the boundaries of science lay (Wright 1981). We would argue that just as in the late seventeenth century – at the beginnings of the institutionalisation of science – in the twenty-first century, the term 'natural knowledge' can be used to open up space for reflection on the complex way in which scientific rationality is constructed, translated and diffused in interaction with human action and everyday knowledge within and outside scientific communities, rather than focusing on the homogeneity of global scientific analysis, which inherently precludes room for reflection on the contours of science in interaction with human agency and the way in which knowledge processes are reciprocally constituted (see Pickering 2001).

Here we explore how some actors connect their scientific understandings to agricultural practices, shaping agricultural change and the repositioning of agricultural science in rural development (see also Cleveland and Soleri, this volume). In so doing we examine how agronomic knowledge becomes part of actors' creative action, dissolving rigid knowledge boundaries through the integration of contingencies and experience (Arce 1989; Arce and Fisher 2003; see also Dove et al. this volume). We do this by presenting four illustrations: the first of an agricultural experiment in the nineteenth century, the second of a chicken-breeding experiment in the early twentieth century and the third and fourth of agricultural research in the late twentieth and early twenty-first centuries. We use the concept of natural knowledge to take forward a reflexive anthropology of development in relation to our understanding of how scientific knowledge is localised and embedded within human action. This brings an actor-oriented perspective on the interplay between contingencies, local situations and knowledge interfaces, helping us to reflect on the importance of understanding scientific procedures in relation to actors' experiences, making associations between people's everyday life and scientific knowledge.

Understanding Natural Knowledge

Historically, natural knowledge has been part of a long and edgy relationship between the specialised knowledge of scientists and the everyday knowledge of people. According to Dear (1990), in the seventeenth century natural philosophers were concerned with the need to establish 'formal scientific structures' to guarantee certain knowledge concerning 'conclusions regarding the way things are or happen in the world' (Dear 1990: 665). Different religious traditions influenced the way natural philosophers acquired and institutionalised 'certain knowledge' and built the early foundations of science (Wright 1981). In France, natural philosophy was rooted in steps established by the Catholic religion in the certification of miracles and the Church's institutional control over this process. In England, Protestantism led to challenges to the traditional monopolies of learning (Church, medicine and the universities); the experimental philosophy of English Protestantism focused on discrete historical events, which made the 'ordinary course of nature' something to be gradually understood through inferential steps rooted in 'matters of fact' (Dear 1990; see also Wright 1981). This history is important because scientific rationality – that is carried in the project of agricultural modernisation, for example – has often been presented as beyond ideological, political and religious struggles, suggesting that the scientific method is outside human passion and belief (for well-established critiques of this view see Wright 1981; 1988; Knorr-Cetina 1981; Latour 1987).

Debates about the relationship between indigenous or traditional and scientific knowledge and the different roles they have to play in people's understandings of the world (see for example chapters by Bodeker, Ellen and Marzano, this volume, or Agrawal 1995; Sillitoe 1998) resonate with this historical view, in so far as questions are raised about how and where claims to knowledge are assembled and who has authority over the process. In this respect it is in the knowledge interfaces between different actors that the sense of making and describing natural knowledge emerges as a capacity of social life and the end result of transactions based on actors' agency, rather than in systemic dichotomies between nature and culture or science and indigenous knowledge (see Arce 1989; Arce and Long 1987, 1992, 1994; Sillitoe 2002; Arce and Fisher 2003). In effect, the social relations necessary to generate these knowledge transactions contribute to the creation of different types of epistemic community that enhance processes of social, economic and cultural differentiation (see Knorr Cetina 1999), while shaping knowledge and new courses of action.

Capturing Knowledge within the Agricultural Research Station

Agricultural research illustrates the way in which scientific practices are experienced and embodied in everyday performances. Here we give two historical examples, which relate to the experimental foundations of agriculture: the development of superphosphates in England (based on AGT 1904; Hall 1920; Usher 1923) and improvements in chicken productivity in the U.S. (based on Cooke 1997).

The development of superphosphates to promote plant growth can be traced to the now famous Rothamsted Agricultural Research Station. In the period 1830–50, the relation of soil minerals to plant growth was an important research concern. A renowned contribution to this research was made by the agriculturalist, John Bennett Lawes, and the chemist, Henry Gilbert, who started to conduct a systematic series of experiments at the Rothamsted property in 1834.

Lawes's and Gilbert's experiments with 'mineral manure' (fertiliser) were inspired by lectures delivered by Justus von Liebig, an analytical chemist, at the Royal Institution. Liebig's work centred on the doctrine of the exhaustibility of soil fertility and the idea that 'plants derive not only their carbon but also their nitrogen from the atmosphere'. Lawes and Gilbert started to test Liebig's doctrine, focusing on 'the putting back into the soil of the ash constituents which the plant has taken from it' (AGT 1904: 171). Lawes thought that Liebig's theory did not correspond to his agricultural experience: however, selecting a number of fields from the estate and using a barn for the purpose of establishing an agricultural laboratory, they began a series of experiments with mineral manure. In the early nineteenth century, the atmosphere had been the focus of attention for understanding plant nutrition: through these experiments, combining Lawes's knowledge of agriculture with Gilbert's knowledge of chemistry, they contributed to a shift in thinking whereby the soil became central to enquiry, helping change the focus of existing theory and proving the value of experimentation on plant nutrition.

Focusing on the soil enabled Lawes and Gilbert to test the effect of different nutrients on a range of agricultural crops, showing how the properties of soil interact with the physiology of plants. They did this by dividing up plots of permanent grassland up and treating them with different quantities of mineral manure; later the hay would be dried, weighed and the plant composition analysed. The success of these experiments led Lawes and Gilbert to patent superphosphate in 1842, alongside which a systematic series of now-famous experiments were started, which continue today. Nevertheless, human experience continued to be an important factor in the application of experimental outcomes. As A.D. Hall, formerly a director of the station at Rothamsted, stated:

> … an account of what is removed from the soil year by year by the crops provides
> very little guidance towards determining the amount of fertilizer which must be
> brought in … it is only by experience, by the knowledge of his own land and the

market conditions which prevail that the individual farmer can tell how high it is profitable for him to farm and therefore to what degree he can utilise the information as to feeding his crops which is provided by field experiments. (Hall 1920: 305–6)

In other words, the 'scientification' of agriculture, important as it was for generating change, could only be incorporated into the everyday practices of farmers through their individual experience: through knowledge of the soil, the market and new scientific advances.

This outlook for mastering and using the logical properties of 'natural knowledge' was based on practical and experimental observations on soil responses to inorganic manures. These well-described pragmatic principles showed the effect of different soil nutrition treatments upon the crop. A number of botanical and pedological properties were better understood, at least in part, from their practical observations. This understanding presented 'independent characteristics' from the knowledge guided by the scientific establishment of institutionalised science represented by Liebig and the Royal Institution.

The other historical example of scientific advance within an experimental station concerns the role of the biologist, Raymond Pearl, in chicken-breeding research in the U.S. between 1907 and 1916 (Cooke 1997). In addition to a background in biology and Mendelian genetics, he had experience of practical farming and relations with farmers working to improve egg production. Pearl's research focused on increasing egg production, with his experiments taking place at a time when the U.S. Department of Agriculture was interested in developing poultry rearing and egg production beyond generating an income for farm women into an industry that could diversify agriculture and make it more profitable.

Pearl realised that farmers had achieved significant success through breeding for decades before the rediscovery of Mendel's work. Scientific work at the time suggested that productivity was based on the hen, so that selection of the best layers, supported by proper feeding, should increase yields. The breeding successes produced by farmers led him to doubt some of the results produced by scientists and to turn instead to examine farmers' breeding practices.

In drawing on farmers' practices, Pearl statistically analysed their breeding data. This led him to question existing experimental conclusions, in particular inaccuracies in data and the need to set up an experiment based on reliable 'trap nests' (a box that locked when a hen entered it and held her captive until the breeder released her) as an essential tool to study the breeding practices of chicken producers. By improved experiments on breeding, Pearl came to realise that selecting the best-laying hens only generated good egg production for a generation but did not improve production beyond. In effect he put in doubt accepted scientific thinking of the time and argued, using evidence from the breeding experiments of farmers, that fecundity was improved through the male line, not the female, and therefore that the cock's fertility was essential for improving egg production.

One of the problems Pearl battled with was how to educate a sceptical farming audience about the value of his work. Although his recommendations had contributed to some limited improvements in agriculture, he realised that his plans did not translate easily into a clear strategy for breeding success. In this respect he was aware that lectures and publications only had limited impact on farmers (Cooke 1997: 84). This led Pearl to emphasise the importance of egg-laying contests as events that aroused wide interest and called attention to the results of good care and breeding. This was in a wider context that emphasised extension and demonstration (see Marzano, this volume), of which egg displays at country fairs became important . The work of Pearl combined genetic theory with practical experimentation and observation of farming practices; this generated improved experimental instruments, and introduced new management practices, in effect changing the practice of poultry rearing and egg production.

The cultural potential of people to act in innovative ways to construct natural knowledge (see Golinski 1998) is related to a set of parameters in which alongside the process of experimentation, people's experience interfaces with scientific principles leading to a new interpretation of the status quo. In these examples, scientific agricultural phenomena became shaped by a negotiation over practices and the significance these practices played in the experimental procedures and protocols through which knowledge was validated and finally accepted as a special agronomic knowledge.

Observations of Transfer of Technology at an Agricultural Research Station in Bolivia

Understanding the rebounding effect of natural knowledge, and the way in which the scientific method encounters politics within a scientific community – with implications for the way in which knowledge is created and applied – can be understood by focusing on the relationships between individuals and the institutional context in which they are located. This example focuses on an agricultural research station, the Bolivian Institute for Agrarian Technology (IBTA) where scientific research is taking place that is associated with the transfer of in vitro technology to local agriculture, in the process seeking to transform local farmers into highly specialised producers of palm products (compare Marzano, this volume, for a case of agricultural research and farmers' practices in rubber development in Sri Lanka). Here the use of science within agricultural development and the subsequent difficulties this generates in relation to local production practices highlights the need to take into account the processes through which knowledge is constructed and transposed, as different actors adopt and transform it within the political context of agricultural development in Bolivia.

While working in Bolivia on a Dutch government programme, colleagues at the Universidad Mayor de San Simon in Cochabamba became aware of our interest in issues of science and agricultural development, especially in El Chapare where the growing of coca is the subject of much political conflict. After a conversation about the effect of international cooperation, a colleague mentioned that the University was conducting biological experiments using in vitro fertilisation. Using overseas development assistance, new laboratories had been built and modern equipment imported that, it was hoped, would make a real contribution to improving farmers' livelihoods.

On arrival at the research station in Chapare we were received by the Director and, after formal presentations, given a tour. The research station has several prefabricated buildings, laboratories built in the last five years to meet its experimental needs and activities. Each accommodates three to five researchers and technicians and their equipment.

In one of the buildings, three women were selecting yucca genetic material, which was stored in a cold room. The supervisor approached and after some discussion told us that he did not agree with the new policy of prioritising resources to research the development of export crops alone. He emphasised that local crops such as yucca and citrus trees should not be forgotten in terms of their importance for research in the tropics. He said that his team has very little equipment and practically the only resources they receive are their salaries. According to him, the scientific contribution of the research station to tropical agriculture was not clear, especially with the importance given to export crops. During the conversation he showed us a laboratory refrigerator, the function of which had been adapted to the circumstances resulting from the neglect of knowledge for local tropical crops: inside were the packed lunches of the team and two cans of Coca Cola. We laughed and he commented that in the tropics refrigerators always have important uses in spite of changing research priorities.

The newest building was donated through international cooperation, and inside the atmosphere and activities were in marked contrast to the previous research unit. Someone explained that they were working on the in vitro biology of palm trees using a gel they make themselves and cell culture facilities provided by foreign aid. He explained that palm products have the potential to transform the economic situation of farmers in the region. The idea was to provide hybrid plants to farmers at a subsided price for a period of two years and, after that, when the export trade of palm was consolidated, they could charge the market price. The supervisor was a young and pleasant man, secure of his mission and proud of his new laboratory. He invited us to look around, but to do so we had to don white coats and wrap plastic around our heads to prevent contamination of the genetic material. Shoes were replaced with sterilised slippers, hands covered by rubber gloves and faces protected by sterilised masks. Inside the facility a line of women and men were conducting cell culture experiments. Centrifuge machines, powerful microscopes, radioisotope instruments, cell incubators, pipettes and hundreds of Petri dishes were in contrast to the shabbiness of the previous tropical crop laboratory. After observing the different processes and the specialised room to store the in vitro palm cells, we returned to unsterilised reality for a brief tour of the nursery.

In the nursery we saw palm plants being raised and prepared for delivery to local farmers. One of the technicians said that biologically engineered palms could intensify the local agricultural system, in the process transforming the region's economy. He added that there was pressure on the research station to distribute palm quickly to the local farmer population and develop a high commitment to in vitro technology in contrast to other less 'modern' biological technologies. In his opinion this was creating some uncomfortable new problems: local farmers did not understand the importance of in vitro technology for agriculture, so they were not happy to accept the 'improved' palm trees. Added to this, new conflicts and rivalry had started to emerge between scientists working in different laboratories within the research station, which was manifest in those working in the old laboratories oriented to local crop production mocking the activities and rigid protocols of people working in the new agro-export laboratories, and those working in the new laboratories having arrogant attitudes towards other researchers and their programmes. Our guide then showed us more palm growing in black polyethylene bags in a greenhouse and others in the experimental fields of the research station. With the observation that these palms were strong and good plants, we ended our tour. He then fell silent before continuing in a confessional voice to offer us his experience of the new situation.

According to him, the establishment of new laboratory practices was a good thing, enabling scientists to practise new skills but what he did not like was the political use of in vitro technology which, he argued, prompted the scientists working with the new biological techniques to criticise research previously undertaken by colleagues using 'old' equipment. According to him this was not fair because new knowledge and technology were used as political weapons within the research station rather than for solving problems in the region. There were also problems with the farmers who did not understand why it was necessary to reproduce plant material in the research station. The idea of a steady and speedy process of palm plant production was hard to integrate into the routine and practices of local farmers, who kept complaining that, as was the case with other crops, there was no reliable market distribution for palms produced. According to our informant, the farmers wanted a better distribution mechanism for the crop and an understanding of the different conditions of tropical soil within the research programme in order reliably to produce the crop, rather than an emphasis on speedy production.

Since the visit in the late 1990s, the development of in vitro technology for crops in the Chapare has virtually halted due to lack of resources to the research station, associated with a slow-down in international cooperation. Hence the national policy of agro-export promotion represented by in vitro palm development was seen as part of a larger policy of coca control and the scientific solution to the region's political and economic problems, while 'older' forms of technology represented past economic conditions and less elaborate scientific practices. Ironically, in the past these 'older' technologies had themselves once been part of a political drive for modernised agricultural development, given that IBTA was set up in 1975 to develop 'alternative' crops to coca production.

When a new scientific technology is introduced into an agricultural research station, it acquires a political value, which is linked not only to the potential to improve existing crops, but also to a series of repositioning acts within the laboratory, changing the cultural repertoire and positions of those working within the research station. In this case attempts to transfer scientific knowledge to farmers sought to orient them towards more specialised understanding of palm production techniques. This brought conflict within the local research establishment and made farmers reluctant recipients of the supplied plant material.

In this example scientific method and expert rationality interact with knowledge practices within the research station, which, rather than generating continuity with existing research and a ready take-up among farmers, serves first and foremost as a vehicle for furthering power relationships between individuals. These power relationships give rise to conflicts and differences that highlight the social nature of agronomic knowledge within a larger domain of development practice and discourse. In this trajectory, biological knowledge, represented by in vitro technology, failed to generate meaningful dialogue between different groups of scientists and between scientists and farmers. Instead the new technology and its representation (for instance, the symbols of advanced science: protective clothing and laboratory paraphernalia) provided a symbolic means for people to see that the nature of agricultural production was changing in the Bolivian tropics.

The Future of Indigenous Agriculture in Contemporary Society

Since biotechnology has become used to create transgenic agricultural crops, there has been a debate concerning potential environmental risks and negative social consequences resulting from these scientific advances. Historically, farming systems have developed from complex relationships between organisms, habitats, ecosystems and the agricultural and cultural practices of farmers. Biotechnology serves to transform these relationships, shifting knowledge boundaries (and farming practices?) in dramatic ways. It has taken us into the field of contested political relations between peasant enterprises, scientists, corporate interests, and governments (see Goodman and Wilkinson 1990; Gremmen 2003). In examining a controversy arising from the attempted patenting of a variety of 'traditional' Andean crop, 'Apelawa' quinoa, we can see the way in which the rebounding of scientific and local knowledge in relation to different actors generates conflict and subsequent changes in social relations vis-à-vis people's property rights.

A glimpse at the controversies that genomics research and development has generated (see Nuffield Council on Bioethics 2004; FAO 2003/4)[3] highlights the contradictions that arise between the interests of farmers and those of institutions such as universities and agrochemical or seed industries. For instance, the biotechnology industry apparently has its power concentrated in the hands of five multinational conglomerates that provide seeds and restrict access to research technologies (see Cleveland and Murray 1997; Roopnaraine 1998). Indeed, it is interesting to note that the United States of America has played a key rôle in constructing the intellectual property regime that legitimises, governs and controls the infrastructure of bio-development, entrepreneurships of Western academia and private funding (for an instructive case that explores issues raised by the U.S. patent system see Clift, this volume).

The consequences of U.S. patent laws can be seen in the case of the Andean crop quinoa, of which there are several different varieties, some of which are increasingly popular in European and North American niche markets (see Laguna 2002a, b). In 1997, Bolivia's National Association of Quinoa Producers (ANAPQUI) filed a claim against two professors at Colorado State University, challenging their patent on a variety of quinoa. The U.S. patent granted the professors exclusive rights over the variety 'Apelawa' quinoa, which is cultivated by peasants near Lake Titicaca. The patent granted rights to use this variety as a source of male sterile cytoplasm in a technique to create hybrid quinoa. This generated controversy over the potential of the patent to rights over any quinoa hybrid derived from Apelawa variety. Some people thought this claim could include rights to old varieties grown by peasants in Bolivia, Peru, Ecuador and Chile, threatening Bolivia's export market in quinoa.

The Colorado professors argued that the patent was restricted to the hybrid varieties generated from the sterile male cytoplasm. This different understanding of the potential of biotechnology processes and claims between the farmers and the professors generated a conflict that continues to the present day. This has resulted in strong opposition to biotechnology amongst quinoa producers.

In counter-arguments to the patent, the ANAPQUI association argued: 'Quinoa has been developed by Andean farmers for millennia; it was not "invented" by researchers in North America'. [Therefore]…'we demand that the patent be dropped and that all countries of the world refuse to recognise its validity'. Bolivian peasants were able to generate global support: they learned how to act against the claim from Colorado University with limited information and uncertainty. Mr Mamani, the president of ANAPQUI, travelled to New York to make an appeal at a special session of the United Nations General Assembly. Then he went on to the International Peoples' Tribunal on Human Rights and the Environment, where he argued that the patent was a violation of human rights. The Bolivians sought the advisory opinion of the International Court of Justice and became active members of the Indigenous Peoples' Biodiversity network and other NGOs (non-governmental organisations).

The mobilisation of the quinoa peasants' organisation gained political strength from the support of NGOs like OXFAM in the U.K. and the Rural Advancement Foundation International (RAFI) in Canada. It was the latter organisation that discovered the patent and informed Bolivian farmers, who were surprised to hear of its existence. As part of a global social counter-movement, local peasants struggled for their rights, but they needed others to act as watchdogs to monitor information and organise global political campaigns against an undemocratic use of the patent regime, in so doing they opposed external control over biotechnology processes involving their crops. Finally, on 1 May 1998, Colorado State University abandoned its application for a U.S. patent. ANAPQUI celebrated this as a victory for Andean farmers. This event was significant for peasants who experienced being part of a global process in which the creation of natural knowledge through biotechnology was energetically challenged by their quinoa organisation. The interplay between NGOs and peasant organisations is noteworthy, which through loosely coordinated actions was able to channel the sentiments of local groups against multinationals and Western institutions.

Controversies concerning intellectual property regimes and rights over indigenous crops can become expressions of value contestations concerning the legitimation of biotechnology and control over the cultural knowledge of indigenous people (see Brown 1998). These controversies are becoming increasingly common in the so-called global knowledge economy, of which genomics research is a recent scientific manifestation. It is apparent that the boundaries between property rights and the potential of native crops for new forms of marketing and consumption have shifted. This coincides with a revalorisation of native crops, a process that has yet to imbue them with their maximum exchange values and therefore, as Strathern (1999) suggests, experimentation with the construction and assembly of new forms of ownership can be seen as part of a new Euro-American form of appropriation of local peasant resources.

Against this background it is understandable why there are political concerns about the growth of the patent regime in both the private and public sectors. The use of patents to claim ownership and control over the transgenic process has created discord amongst the bigger corporations but has also led to major conflicts of interest between multinationals, academics, and small peasant farmers in developing countries. Access to plant genetic resources is critical for development and research of genetically modified crops, and legal tussles have emerged over the appropriation of local knowledge of crop varieties to this end and their monopolisation using Western patent regimes. These conflicts are acute in relation to control over local staple crops.

As we have tried to suggest, agricultural knowledge and social change are linked to the make-up of social relations, which are increasingly influenced by advances in scientific knowledge. These advances incorporate scientific explanation and a realisation of the potential held by new sources of natural knowledge as they become embedded within existing social practices. This can create an impression of the certainty of scientific knowledge in providing unique inter-pretations of everyday problems. However, as the defining characteristics of soci-ety change and new characteristics emerge, scientists, farmers and others chal-lenge existing interpretations of scientific knowledge and everyday practices. Since knowledge depends on an infinity of possible choices and open problems, it is public debate and opinion formation on value -based issues that limits it via the images of what is legal, ethical, modern or traditional, which determines choices. This being so, natural knowledge is constantly redefined and redirected as it has a bearing on novel questions arising in society.

Conclusion

In this chapter we have used the concept of natural knowledge to consider how scientific knowledge is transformed by creative action, power relations and inter-action within different knowledge frames. Thus, in the cases of soil experiments in the nineteenth century and chicken-breeding experiments in the early twen-tieth century scientific understandings were taken forward through experimen-tation linked to creative action, bringing together farmers' practices, agricultural techniques and experiential knowledge. In the case of the research station in Bolivia, issues of power and the political value of knowledge come to the fore within a research community and in the interactions between scientists and local farmers. In the final case, while there is clearly an unequal distribution of power between local farmers, international scientists and other actors, natural knowl-edge emerges in the capacity of contemporary science to shift from a biological orientation to a biotechnological one, depending on information arrangements and global support specific to any local situation.

Emerging from the interface between the outline of natural knowledge and the form of agriculture, science and experiments, the pragmatic regime of every-day engagement generates analytical constructs for an anthropology of develop-ment concerned with realising processes of reflection. This underlines the way that processes of knowledge creation and transformation can go beyond science, while offering a picture of contradictory situations guided by scientific knowl-edge in the construction of modern life and global processes.

Notes

1.   The Royal Society of London for the Improving of Natural Knowledge (Wright 1981: 97).
2.   By the relationship between human and non-human elements we mean the heterogeneous association between human social life, machines and artefacts (see Golinski 1998: 41; also the work of Haraway 1991; Latour 2004).
3.   See the open letter from non-governmental organisations to the FAO Director General, June 2004, and his reply in the Biotechnology section of the FAO website.

References

Agrawal, A. 1995. 'Dismantling the Divide between Indigenous and Scientific Knowledge', *Development and Change* 26: 413–39.

AGT. 1904. 'The Rothamsted Agricultural Experiments', *New Phytologist*, 3(6/7): 171–76.

Arce, A. 1989. 'The Social Construction of Agrarian Development: a Case Study of Producer-Bureaucrat Relations in an Irrigation Unit in Western Mexico', in *Encounters at the Interface: A Perspective on Social Discontinuities in Rural Development*, (eds). N., Long pp. 11–52. Wageningen University Press: Wageningen.

Arce, A. and E. Fisher 2003. 'Knowledge Interfaces and Practices of Negotiation: Cases from a Women's Group in Bolivia and an Oil Refinery in the United Kingdom', in *Negotiating Local Knowledge: Power and Identity in Development*, in, (ed.) J. Pottier, A. Bicker and P. Sillitoe, pp. 74–97. Pluto Press: London.

Arce, A. and N. Long 1987. 'The Dynamics of Knowledge Interfaces between Mexican Agricultural Bureaucrats and Peasants: a Case Study from Jalisco, Boletin de Estudios Latinoamericanos y del Caribe', *CEDLA* 43, December: 5–30.

———— 1992. 'The Dynamics of Knowledge Interfaces between Bureaucrats and Peasants', in *Battlefields of Knowledge; the Interlocking of Theory and Practice in Social Research and Development*, (eds), N. Long and A. Long, pp. 211–46. London and New York: Routledge.

———— 1994. 'Re-positioning Knowledge in the Study of Rural Development', in *Agricultural Restructuring and Rural Change in Europe*, (eds), D. Symes and A. Jansen, pp. 75–86. Wageningen Sociologische Studies WSS 37, The Netherlands.

———— 2000. (ed.) *Anthropology, Development and Modernities: Exploring Discourses, Counter-Tendencies and Violence*. London and New York: Routledge.

Brown, M. 1998. 'Can Culture be Copyrighted?', *Current Anthropology* 39(2): 193–222.

Cleveland, D. and S. Murray 1997. 'The World's Crop Genetic Resources and the Rights of Indigenous Farmers', *Current Anthropology*, 38(4): 477–515.

Cooke, K.J. 1997. 'From Science to Practice, or Practice to Science? Chickens and Eggs in Raymond Pearl's Agricultural Breeding Research, 1907–1916', *Isis*, 88(1): 62–86.

Dear, P. 1990. 'Miracles, Experiments, and the Ordinary Course of Nature', *Isis*, 81(4): 663–83.

Ellen, R. and H. Harris 2000. 'Introduction', in *Indigenous Environmental Knowledge and Its Transformations: Critical Anthropological Perspectives*, (eds), R. Ellen, P. Parkes and A. Bicker, pp. 1–34. The Netherlands: Harwood Academic Publishers.

FAO 2003/4 'Agricultural Biotechnology Meeting the Needs of the Poor?', in *The State of Food and Agriculture*. Rome.

Golinski, J. 1998. *Making Natural Knowledge: Constructivism and the History of Science*. Cambridge: Cambridge University Press.

Gremmen, B. 2003. 'Towards Worldwide Sustainable Food Security?', in *Genes For Your Food–Food For Your Genes. Societal Issues and Dilemmas in Food Genomics*, (eds), R van Est, L. Hanssen and O. Crapels, pp. 45–56. Rathenau Institute, The Hague. Working Document 92.

Goodman, D and J. Wilkinson 1990. 'Patterns of Research and Innovation in the Modern Agro-Food Systems', in *Technological Change and the Rural Environment*, (eds), P. Lowe, T. Marsden and S. Whatmore, pp. 127–46. London: David Fulton Publishers.

Hall, A.D. 1920. *Fertilizers and Manure*. New York (publisher not documented).

Haraway, D. 1991. 'A Cyborg Manifesto: Science, Technology and Socialist-Feminism in the Late Twentieth Century', in *Simians, Cyborgs, and Women: the Reinvention of Nature*, pp. 149–81. New York: Routledge.

Knorr Cetina, K. 1981. 'Introduction: the Micro-sociological Challenge of Macro-sociology: towards a Reconstruction of Social Theory and Methodology', in *Advances in Social Theory and Methodology towards an Integration of Micro-and Macro-sociologies*, (eds), K. Knorr-Cetina and A.V. Cicourel. Boston, London and Henley: Routledge and Kegan Paul.

———— 1999. *Epistemic Cultures: How the Sciences Make Knowledge*. Cambridge, MA: Harvard University Press.

Laguna P. 2002a. 'Competitividad, externalidades e internalidades, un reto para las organizaciones económicas campesinas: La inserción de la Asociación Nacional de Productores de Quinua en el mercado mundial de la quinua', in *Debate Agrario*, no. 34, Centro Peruano de Estudios Sociales (CEPES), Lima, pp. 95–169.

———— 2002b. 'La Cadena Global de la Quinua: Un Reto para la Asociación Nacional de Productores de Quinua', in *La gestión económica-ambiental en las cadenas globales de mercancías en Bolivia*, (eds), C. Romero Padilla and W. Pelupessy. pp. 89–195. IESE-PROMEC-IVO, Tilburg University and Universidad Mayor de San Simón, Cochabamba, Bolivia.

Latour, B. 1987. *Science in Action*. Stony Stratford: Open University Press.

———— 1988. The Pasteurization of France, trans. Alan Sheridan. Cambridge, MA: Harvard University Press.

———— 2004. *Politics of Nature: How to Bring the Sciences to Democracy*. Cambridge, MA: Harvard University Press.

Nuffield Council on Bioethics 2004. *The Use of Genetically Modified Crops in Developing Countries: A Follow-up Discussion Paper*. 121 pp. Members of the working group: S. Thomas, D. Burke, M. Gales, M. Lipton and A. Weale.

Parkin, D. 1995. 'Latticed Knowledge: Eradication and Dispersal of the Unpalatable in Islam, Medicine and Anthropological Theory', in

    Counterworks: Managing the Diversity of Knowledge, (ed.) F. Fardon, pp. 143–63. London and New York: Routledge.

Pickering, A. 2001. 'Practice and Posthumanism: Social Theory and a History of Agency', in *The Practice Turn in Contemporary Theory,* (eds), T.R. Schatzki, K. Knorr Cetina and E. Von Savigny, pp. 163–74. London and New York: Routledge.

Roopnaraine, T. 1998. 'Indigenous Knowledge, Biodiversity and Rights', *Anthropology Today* 14(3): 16.

Richards, P. 1985. *Indigenous Agricultural Revolution.* London: Hutchinson.

Sillitoe, P. 1998. 'The Development of Indigenous Knowledge: a New Applied Anthropology', *Current Anthropology.* 39(2): 223–52.

Sillitoe, P. 2002 'Globalizing Indigenous Knowledge', in *Participating In Development: Approaches To Indigenous Knowledge,* (eds), P. Sillitoe, A. Bicker and J. Pottier, pp. 108–38. London: Routledge.

Strathern, M. 1999. *Property, Substance and Effect: Anthropological Essays on Persons and Things.* London and New Brunswick, NJ: The Athlone Press..

Usher, A.P. 1923. 'Soil Fertility, Soil Exhaustion, and Their Historical Significance', *The Quarterly Journal of Economics* 37(3): 385–411.

Warren, M., J. Slikkerveer and D. Brokensha (eds) 1995. *The Cultural Dimension of Development: Indigenous Knowledge Systems.* London: Intermediate Technology Publications.

Wright, P.W.G. 1981. 'On the Boundaries of Science in Seventeenth-century England', in *Sciences and Cultures: Anthropological and Historical Studies of the Sciences,* ed, E. Mendelsohn and Y. Elkana pp. 77–100. Dordrecht, Boston and London: D. Reidel Publishing Company.

10 Is Intellectual Property Protection a Good Idea?

Charles Clift[1]

The issue raised in this chapter does not relate to the debate about the scientific status of traditional knowledge in the eyes of the 'modern' scientific tradition,[2] or vice versa.[3] It is assumed, as also seems widely accepted, that traditional knowledge has an important role to play in contributing to the progress of science, and to its application to improve human health and welfare. Traditional knowledge is complementary rather than competitive with modern science. Each has something to learn from the other, as argued by several other contributors to this volume. The holistic and systemic worldview of traditional knowledge, which emphasises the interconnectedness of both natural and human phenomena, often provides fresh insights to modern science. The latter, which tends to proceed by isolationist and reductionist methodologies, may miss the bigger picture which traditional knowledge offers.

The topic addressed here is whether there is anything to be learnt from the application of intellectual property rights (IPRs) to modern science in universities or public research institutes, in the context of the debate about whether a form of intellectual property (IP) protection should be introduced for traditional knowledge. The nature of the innovation process in traditional knowledge and modern science may be different, incorporating divergent but complementary worldviews (e.g. see ICSU 2002: 11), but in both cases the progression of knowledge has not relied on the incentives provided by the intellectual property system. Can the debate on IP and traditional knowledge be informed by the experience of the application of IP in university-based modern science in the last two decades or so?

In the case of modern academic science, the knowledge produced has traditionally been put into the public domain through publication or other means of dissemination, free for anyone to use, for commercial gain or not. However, in recent years, governments have increasingly encouraged public research institutions to take out patents on the technologies they produce on the grounds that this facilitates their further development and their commercial application, and thus promotes social and economic welfare. This means that, although the knowledge may still be put into the public domain through disclosure in patent applications, the freedom to use it is restricted to those the patent holder chooses to license, and by the cost of the licence. Moreover, it gives the research institution, and also scientists within it, a financial interest in maximising the amount of patenting, and thereby royalties from licensing. This may create a conflict with the traditional function of widely disseminating new knowledge and techniques so that all may benefit, and build upon them.

In the case of traditional knowledge, the status of the knowledge produced is less clear. It may be confined to an individual, a few people in a particular community, or it may be widely known throughout a community or outside it. At the other extreme, it may be published as, for instance, in the ancient Indian medical texts such as the Ayurveda, and be wholly in the public domain (see Bodeker, this volume). However, like academic science, the motive for producing knowledge has traditionally not been commercial. Academic scientists are paid, inter alia, to produce new knowledge but the incentives for advancing knowledge relate more to prestige, peer recognition and career advancement than direct financial incentives. Traditional knowledge is similar in the sense that it has developed organically in response to the livelihood and health needs of local communities, not through any institutionally defined property rights in knowledge.

Just as intellectual property protection has recently been extended to the realm of academic science, so also there is an active debate about how the intellectual property system should be modified or extended to accommodate or to protect traditional knowledge. While the parallel is not always very exact, there are sufficient similarities in these two processes to consider whether we can learn from the experience of patenting in the academic world in extending intellectual property protection to traditional knowledge.

Traditional Knowledge and Scientific Knowledge

Academic scientists share with indigenous communities a set of rules, institutions and social norms which, however different, appear to be conducive to the growth of knowledge. Such norms include the emphasis on publication as the route to promotion and prestige, the peer review process, the ritual of citation, the balance between cooperative and competitive behaviour, the extent of networking activities to share knowledge formally and informally and so on. These norms of 'open science' are 'functionally quite well suited to the goal of maximising the long-run growth of the stock of scientific knowledge' (Dasgupta and David 1994: 518). The incentives are for the most part not financial, but based on intellectual curiosity, peer recognition and competition to be first in advancing the frontiers of knowledge. In this world, scientists are paid by the state (or other benefactors) to pursue scientific endeavour where the overwhelming emphasis is on enlarging the size of the public domain of knowledge. Enlarging this domain allows the maximum benefit to be derived by exposing new knowledge to the largest possible audience.

One can imagine that systems that result in the accumulation of traditional knowledge have different but analogous community and individual mechanisms that also function to advance knowledge. The rituals and institutions responsible for the advance of knowledge are undoubtedly very different. But both are, or were, fields where the motive for the advancement of knowledge was not the financial incentive offered by the intellectual property system, or the privatisation of knowledge by individuals, communities or companies.

Both sets of knowledge are for the most part put in the public domain. In the case of both traditional knowledge and modern science there may be exceptions to this. Individuals and communities may keep knowledge secret for a whole variety of reasons (religion, competition with others), just as scientists may not publish knowledge, or may keep it in their immediate circle, in order to retain some competitive edge over others in the same field. But there is some presumption that use of knowledge within the community, if not always outside it, should not be impeded.[4]

It may be significant that, in both cases, the knowledge that is principally of value relates to the life sciences. Traditional knowledge and associated genetic resources are overwhelmingly related to either medicine or agriculture. In the U.S.A., approximately half of patents obtained by universities are in the bio-medical area, and the proportion of licensing income derived from this area is considerably higher (Rai and Eisenberg 2003: 292).

The parallels between traditional knowledge and modern science should not be overstated. In particular, the state (or other benefactors) pays scientists to pursue their trade, whereas traditional knowledge arises as an integral part of the search for a better livelihood by (often poor) communities. In that sense, there is indeed a moral case for remuneration when that knowledge is used for commercial purposes. The question is the extent to which the application of IPRs in both these fields contributes principally to the public good rather than private gain.

The Experience of Patenting in Academia

In 1980, the U.S. passed the Bayh–Dole Act, which permitted universities to acquire intellectual property rights for technologies produced by federally-funded research in order to facilitate the development and application of such technologies, by the private sector (known as 'transfer of technology'). Since then other countries, particularly in the developed world but increasingly in the developing world, have adopted similar policies. For the first time the ownership and licensing of intellectual property rights has become a prominent issue for academic scientists.

Scientists, brought up in their 'traditional' culture, have generally reacted warily to the introduction of the IPR system into the structure of university incentives. The U.K. Royal Society in a report in 2003 concluded:

> In short, although IPRs are needed to stimulate innovation and investment, commercial forces are leading in some areas to legislation and case law that unreasonably and unnecessarily restrict freedom to access and use information and to carry out research. This restriction of the commons by patents, copyright and databases is not in the interests of society and unduly hampers scientific endeavour. (Royal Society 2003: 6)

On the other hand, there are strong advocates for the beneficial impact of the system. The director of the Technology Licensing Office at the Massachusetts Institute of Technology (MIT) wrote:

> the impact of university technology transfer on the local and national economies has been substantial, and leads to the conclusion that the Bayh–Dole Act is one of the most successful pieces of economic development and job-creation legislation in recent history. It has been estimated that more than 200,000 jobs have been created in the United States in product development and manufacturing of products from university licenses, with the number increasing fairly rapidly as the licenses mature. (Nelsen 1998)[5]

It is difficult to isolate the impact of Bayh–Dole from the influence of concomitant changes in technology (notably the rise of biotechnology) (Mowery et al. 2001) and in patent policy generally (the advent of the 'pro-patent era' in U.S. policy) (Jaffe 2000). In general, the academic evidence from the United States suggests that the rapid increase in patenting since the late 1970s, which followed the passing of the Bayh–Dole Act in 1980, has not had the specific impact some claim in accelerating the application of scientific knowledge developed in universities. Nor has it had any significant financial benefit for universities overall, despite one or two cases where universities have licensed very lucrative technologies. Most of the income from licensing patents is eaten up by the costs of running technology transfer offices, maintaining the patent portfolio and of legal services.[6]

Moreover, there is evidence that it can lead to internal conflicts in universities and conflicts of interest that damage the collegial culture. It can complicate doing research by restricting access to important technologies that might be freely available in the absence of patenting. It threatens the rationale for public support of scientific research. Above all, the case for asserting as a generality that patenting accelerates the transfer of technology and its application, as compared to putting research results directly into the public domain, is not proven.[7]

This is not to say that the case for patenting university technologies is absent. That case rests on providing the incentive for the private sector to invest in the development of technologies to a state where they can be commercially applied. For particular technologies, exclusive licensing to a suitable company may be the optimal route. For instance, a new compound that needs large-scale investment to determine its efficacy as a drug and the subsequent testing, trials and manufacture would be a suitable candidate for an exclusive licence to a pharmaceutical company. No company might otherwise be prepared to take the risk involved in making the necessary investment.

But the evidence suggests that, once a technology transfer office is established in a university, the financial motive can tend to take precedence over the technology transfer objective. The initial report of the Lambert Review of Business–University Collaboration in the U.K. noted that:

> The consultation revealed that many universities see revenue generation as one of the main objectives of technology transfer. This is despite clear evidence from the U.S. that even the most successful universities only earn small sums from such activities, while many do not manage to break even. Several U.S. universities explained that their main goal was to move technology to the private sector, while revenue generation was seen as a secondary objective. (Lambert 2003: 16)[8]

If the objective becomes the maximisation of income from patenting and licensing, this will tend to conflict with the objective of promoting the application of technologies for the benefit of consumers. In particular, it encourages the patenting and licensing of 'upstream' research products (e.g., diagnostic tools, or research techniques, or the use of genes for diagnostic purposes) that are widely required by other researchers in universities or companies. Patenting and licensing these kinds of technologies on an exclusive basis is likely to militate against the technology transfer objective. Licensing them on a non-exclusive basis can simply amount to a tax on other users of this technology and thus hinder rather than help the eventual development of products. The purposes of the Bayh–Dole Act might be better served by not patenting in these circumstances:

> The farther removed university-based research discoveries are from end-product development, the more likely it is that subsequent research will generate additional patents that will be more important to the profit expectations of private investors than patents on the prior research. Indeed, patents on the many discoveries that enable product development are more likely to add to a product's costs than to enhance its profitability. Given that the long course of biopharmaceutical product development typically generates a great many patented inventions on the road to market, Congress' fear that potential new products would never be developed if the early discoveries from which they sprang remained unpatented seems quaintly out of touch with contemporary R&D and patenting practices. (Rai and Eisenberg 2003: 301–2)

Indeed, the U.S. National Institutes of Health, the largest public-sector funder of health research in the world, with a budget now approaching $30 billion per annum, has recently published 'Best Practices for the Licensing of Genomic Inventions', which reinforce this point:

> Therefore, in considering whether to seek patent protection on genomic inventions, institutional officials should consider whether significant further research and development by the private sector is required to bring the invention to practical and commercial application. Intellectual property protection should be sought when it is clear that private sector investment will be necessary to develop and make the invention widely available. By contrast, when significant further research and development investment is not required, such as with many research material and research tool technologies, best practices dictate that patent protection rarely should be sought. (Department of Health and Human Services 2004)

Moreover patenting can perversely damage relationships with industry through arguments over the price of university-developed technologies. The Lambert Review quoted one industry respondent as saying: 'We have walked away from some university research contracts in the U.K. because the demands on intellectual property were both unreasonable and unrealistic.' Conversely, a university respondent said: 'Many [U.K. companies] expect that they can pay under the odds for the research yet acquire ownership of all the results.' Another business respondent said: 'We have been dismayed at the excessive waste of time, imagination and money spent in arguing over the fine details of formal agreements and ownership of IP' (Lambert 2003: 14–15). The scope for conflict, and of high transaction costs in funding and conducting research, is evident.

Intellectual Property and Traditional Knowledge

The debate on the protection of traditional knowledge, and the genetic resources on which much traditional knowledge is based, was to a large extent triggered by Article 8j of the Convention on Biological Diversity (CBD) which was agreed in Rio in 1992:

> Members should respect, preserve and maintain knowledge, innovations and practices of indigenous and local communities embodying traditional lifestyles relevant for the conservation and sustainable use of biological diversity and promote their wider application with the approval and involvement of the holders of such knowledge, innovations and practices and encourage the equitable sharing of the benefits arising from the utilization of such knowledge, innovations and practices. (Convention on Biological Diversity 1992)

The language asks signatories to 'respect, preserve and maintain' traditional knowledge, to 'promote [its] wider application' and 'encourage the equitable sharing of the benefits' arising from its utilisation. It thus seems to encourage countries to introduce some form of IP-like protection for traditional knowledge which would also promote its wider application and ensure benefits are shared with the holders of traditional knowledge.

It also implicitly suggests that the application of the current intellectual property system may not be in the interests of the originators of traditional knowledge. This reflects the concern that firms or individuals may 'steal' traditional knowledge and privatise it by applying for patents on it. This is wrong because it is theft of people's 'intellectual property' and may also contravene moral, spiritual or cultural mores. For instance, a plant may be considered sacred and its patenting and commercialisation may deeply offend (e.g., the ayahuasca plant from the Amazon, which was patented in the U.S.A.) (Commission on Intellectual Property Rights 2002: 76–77).

Moreover, patenting may have an economic effect. Most commonly, there is the risk that exports of the product or knowledge now patented will be prevented (e.g. to the U.S. or Europe). This was, for instance, a concern of India in the patents relating to Basmati rice (Commission on Intellectual Property Rights 2002: 89), and in the case of the patent on the Enola bean, where the U.S. patentee has attempted to sue Mexican companies importing yellow beans into the U.S. (Rattray 2002). In any event, there is the concern that outside parties use the patent system to capture for themselves the commercial rewards from the invention, without any obligation to share those rewards with its originators.

In some cases the principal concern is that a patent has been wrongly awarded. If the traditional knowledge is in the public domain (or constitutes 'prior art' in patent jargon), then a patent should not be awarded because it would fail to meet the criteria for patentability relating to novelty (the invention is not new) or inventiveness (it adds nothing to what is already known or published). For example, the Indian government managed to overturn claims in a U.S. patent relating to the wound-healing properties of turmeric by bringing the 'prior art' to the attention of the U.S. Patent and Trademark Office (Commission on Intellectual Property Rights 2002: 76).

In other cases, the patent may be correctly awarded because the claimed invention, while drawing on the traditional knowledge, sufficiently adds to it to meet legitimately the patentability criteria. For instance, the South African Council for Scientific and Industrial Research patented the appetite-suppressing component of the Hoodia plant which they had isolated following investigations of the traditional use of the plant for this purpose by the San people of southern Africa. In this case, the issue debated was not whether patenting was invalid or wrong per se, but whether suitable arrangements had been made to provide the San with a share of the possible benefits of commercialisation of the invention derived from their traditional knowledge (Commission on Intellectual Property Rights 2002: 77. See also Sillitoe, this volume).

Peculiarities in national patent legislation may also allow patents to be granted which would not be awarded elsewhere. The principal example here is the U.S.A., which ignores undocumented 'prior art' overseas. Thus India could overturn the turmeric patent in the U.S. because it was able to produce published documents demonstrating knowledge of turmeric's wound-healing properties. Had that knowledge been simply oral, as is often the case with traditional knowledge, it is unlikely India could have overturned the patent.

Just as there is a wide disparity in the issues that raise concern, so also are there a number of schools of thought about how they might be addressed. Suggestions or actions include:

- Development of publicly accessible databases of traditional knowledge as 'prior art' to prevent the issue of invalid patents, e.g. as for turmeric.
- *Sui generis* national legislation on access to genetic resources and benefit sharing to implement the CBD principles.
- The development of international rules, perhaps leading to a treaty, to protect traditional knowledge and combat 'biopiracy' (which normally involves more than one territorial jurisdiction and seems therefore to require international action).
- Introducing a requirement (e.g. in TRIPS, the World Trade Organization's Agreement on Trade-Related Aspects of Intellectual Property Rights) obliging disclosure in patent applications of the geographic source of origin of genetic resources from which the claimed invention is derived. Beyond that, many suggest as a condition of patentability evidence that national legislation on access has been adhered to, such as prior informed consent of traditional knowledge holders and benefit sharing.
- Making traditional knowledge unpatentable as a means of protecting it from commercialisation, and protecting the way of life of indigenous or other communities for whom it is important.
- Introducing a new category of IPRs to protect traditional knowledge and facilitate its commercialisation.

These are a wide variety of solutions to the problems identified, some of them contradictory. Underlying this somewhat confused situation is a lack of clarity about the ultimate objective of the 'protection' of traditional knowledge. Many traditional knowledge holders may not wish their knowledge to be commercialised, or to make money out of it – they do not regard it as property to be bought or sold. They may want to 'protect' it from wider disclosure and from commercialisation. Others may want exactly the opposite: establishing a fair return from wider exposure and commercial exploitation. Often, the whole argument about traditional knowledge is symptomatic of much bigger questions relating to the position of communities within the wider economy and society, and their access to or ownership of land and other resources. Thus the 'protection' of traditional knowledge may be an unsuitable policy instrument for resolving the more fundamental problems affecting communities, or even a diversion from addressing them more directly. One non-governmental organisation in this field contends:

> Framing people's rights to traditional knowledge as intellectual property rights is simply wrong ... The privatisation and commercial appropriation of traditional knowledge through intellectual property rights is one of the major *threats* [italics in original] to traditional knowledge systems, not a route to safeguarding them. (GRAIN 2003: 4)

Traditional Knowledge Protection and Public Policy

The essential public policy purpose of the patent system is to overcome a market failure which may cause private firms or individuals to invest too little in research and development. The patent system, by providing a temporary monopoly on the use of inventions, provides that incentive and also obliges public disclosure of the nature of the invention. The cost of the system to users of the patented invention is that they pay more than they would have for a similar invention funded otherwise (e.g., by public funding). Of course, the nature of the system, which works by conferring a private right, means that its users, the patent holders, have a commercial interest in the strengthening and extension of the system. But the concern of policy makers should be to balance 'the public interest in accessing new knowledge and the products of new knowledge, with the public interest in stimulating invention and creation which produces the new knowledge and products on which material and cultural progress may depend'(Commission on Intellectual Property Rights 2002: 6). Interest groups associated with rights holders, often influential, may seek to shift the balance too far beyond the level of incentives needed to promote invention and creation. By contrast, consumer interests, although often less influential, may prioritise current access over future invention.

Thus, the principal justification for the patent system relates to the need to provide an incentive for invention and innovation, in the interests of the wider society. Although the system also has distributional consequences (patent holders reap income from consumers), this is an outcome of the system, not a reason for its existence. This is not to say that the distributional aspects are unimportant. On the contrary, particularly in developing countries, the distributional consequences often appear perverse, which is why, for instance, the relationship between drug pricing and the patent system is so controversial.

As applied to traditional knowledge, the major arguments for protecting it relate to arguments based on equity, rather than incentives. As is evident, traditional knowledge as it has evolved to date has done so in the absence of intellectual property protection. It is therefore impossible to argue that IP protection is necessary to maintain that evolution as the pharmaceutical industry, for instance, would certainly argue is the case with new drug discovery. The nearest approach to an incentives argument is that because many communities are under threats of various kinds (social, cultural, political and economic) their knowledge is in danger of being lost and that 'protection' would help to preserve knowledge that might otherwise disappear. The wording of the CBD ('respect, preserve and maintain') supports that interpretation.

But this argument partly rests on a semantic illusion. 'Protection', in the intellectual property sense, refers to the assignment of rights (e.g., to exclude others from utilising an invention) and has a different meaning from 'protection' of traditional knowledge, where this refers to avoiding the loss of such knowledge as a result of wider economic and social forces. The latter use of the word is really a synonym for 'safeguarding'. There is not an obvious logical causal link which leads from 'protection' in the intellectual property sense to 'safeguarding' traditional knowledge against loss.

Generally speaking, the argument used is that where knowledge developed for communities' own purposes generates value for the wider world, then communities should be compensated appropriately for their part in creating this knowledge. This is an argument about fairness or equity, not about incentives for invention and creation. A recent paper submitted to the TRIPS Council by the African Group expressed this well:

> The protection of genetic resources and traditional knowledge ... is an important means of addressing poverty and is rightly a matter of equity and due recognition for the custodians of the genetic resources and traditional knowledge. It is also a matter of law in the context of protecting cultural rights as well as of preserving the invaluable heritage of humankind that biological diversity and traditional knowledge constitute. (WTO 2003: 1–2)

Thus the grounds for regarding traditional knowledge as itself worthy of intellectual property protection are questionable. If, in fact, the central policy concern is about the poverty of communities with traditional knowledge and/or the preservation of their heritage and cultural values, it is likely that other measures, mainly outside the sphere of intellectual property, may be far more effective at achieving those policy objectives. Another way of putting this is to ask where intellectual property protection would be likely to feature in the development priorities of indigenous groups or local communities. The answer is probably very considerably lower than the priority accorded to the issue in international debates.

Moreover, measures outside the sphere of intellectual property may avoid the costs that intellectual property protection of traditional knowledge would impose on the wider world. For instance, the more extreme versions of protection might give exclusive rights in perpetuity to some kinds of traditional medicine or crop varieties. The effect of such protection would be to raise the cost of these items and reduce the access of consumers, without even the offsetting benefit of an incentive to innovation – indeed quite possibly erecting a further barrier to innovation based on these protected resources. Traditional knowledge that was previously in the public domain and therefore available for anyone to make use of, might now have restricted access.

Conceptually, access- and benefit-sharing regimes (as have been introduced in a few countries)[9] may be distinguished from the conferring of intellectual property protection, although the latter is often seen as facilitating the former as some form of right is necessary to commercialisation and the creation of benefits to be shared. The proposal that applicants for patents should be obliged to disclose the source of origin (as noted above) makes use of this connection.

But the mechanics of establishing operational access- and benefit-sharing regimes and demonstrating their benefits has proved problematic. First, the procedures for establishing prior informed consent and the details of benefit-sharing arrangements are not well developed and have high transactions costs, in particular because the communities may be dispersed and not very cohesive. Secondly, the costs of putting such arrangements into effect, which seem quite high, need to be compared with the potential value of the underlying resource, to the community or to the world at large. The evidence to date suggests that the potential value of genetic resources is very much less than originally envisaged. In particular, communities may well have inflated expectations about the value of their genetic resources or traditional knowledge, which may lead them to withhold access to third parties who wish either to do scientific research or to bioprospect for potential pharmaceutical compounds or treatments. Thirdly, there is a danger that contractual arrangements, for instance in the form of Material Transfer Agreements (MTAs), will place restrictions on the subsequent use to which genetic resources materials are put, rather as patents on 'upstream' university technologies may restrict further research, as noted above.

From countries that have introduced access- and benefit-sharing regimes there is already evidence of how such regimes may restrict legitimate access on the part not just of commercial companies but also of scientists seeking to undertake legitimate fundamental research. For instance, scientists in Brazil have urged the government to modify the laws introduced to curb biopiracy because they are stifling their research on Brazilian biodiversity.[10] An analysis in the *New York Times* in 2002, reviewing similar concerns in a number of countries, attributed this to the national laws and policies put in place to implement the CBD. The CBD was based on the premise that rainforests would prove the lucrative 'medicine chests for the world'. It quotes John Barton, of Stanford University, as saying that the CBD 'is much more about sharing the profits from genetic resources than it is about conserving biodiversity, about science' (Revkin 2002). A university scientist from Columbia, after being denied meaningful access to a flowering plant he wanted to research in a remote part of Columbia, said:

> There is no way that our societies in Latin America will emerge from centuries of poverty while holding a completely distorted view of nature. Once we start looking at organisms as bank accounts, then we are missing the entire view of what is in front of us. Curiosity of the living world ends and so does the meaning of being here. (Agres 2003)

There is also evidence of a threat to the conservation of genetic resources ex situ and to future research that depends on such collections. An adviser to the International Plant Genetic Resources Institute (IPGRI), one of the research institutes of the Consultative Group on International Agricultural Research, which hold large collections of genetic material for crop breeding and research relevant to developing countries, noted:

Our task has become extremely difficult since passage of the CBD. A lot of material in our gene bank would be extinct if it had not been collected in the past. But
we're finding it harder and harder to collect materials now. Collection samples have
dropped from about 30000 per year to fewer than 5000 as a direct result of the
treaty. The CBD is both the cause and effect of this mentality. For years, it was sac-
rilegious to say anything against the CBD. If you did, you were reactionary and
anti-developing countries. But at what point do you say the emperor has no clothes?
The facts do not support this treaty as being terribly productive. (Agres 2003)

The argument about the protection of traditional knowledge, and of regulations
on access and benefit sharing, largely ignores the context in which other new
forms of intellectual property protection are discussed. Many recent or proposed
extensions of intellectual property protection (e.g. the protection of databases in
the EU, or the application of copyright to the digital arena) are controversial, as
in the case of traditional knowledge. But their logical context is, or should be,
the perceived need to provide or maintain incentives for creative activities which
are judged to be in the public interest, not (at least so explicitly) about the material advantage of the potential providers of knowledge or creators. In the case of
traditional knowledge, this logic is turned on its head. The money appears to be
more important than the conservation, or the incentive to research on new crops
or medicines.

The controversial nature of extensions of intellectual property rights often
occurs because critics judge that the balance between the public interest for protection, and the material interests of those protected has shifted too far in favour
of the latter. In most cases, also, the latter are not poor, and may often be large
multinational businesses. However, because communities that possess traditional knowledge are generally poor in material terms, it is judged politically incorrect to criticise proposals for its protection on the grounds that, although protection may benefit knowledge holders, it does little or nothing to promote the
wider public interest by stimulating innovation or encouraging the application
of knowledge.

It is an irony of this debate that the advocates of protection for traditional
knowledge are often those who have criticised the imposition of intellectual property regimes, and of new intellectual property rules, on developing countries and
the increasing 'privatisation' of knowledge. A critical issue, which is often ignored,
is the value of the 'public domain' as a source of knowledge and new inventions.
Giving exclusive protective rights to traditional knowledge threatens the availability of this knowledge for the greater good. Moreover, by assigning quasi-own-
ership rights, it may disrupt community processes that contribute to the continuing advance of traditional knowledge, thus defeating a central purpose of protection. It may contradict the practices and values of some communities.

In that sense, the extension of intellectual property rights to traditional
knowledge needs to be subjected to the same critical analysis as the extension of
intellectual property rights in other fields. In many ways there are close parallels

with the way IPRs have entered into the territory of academic science in the last twenty years or so.

From our analysis above of the impact of Bayh–Dole we can draw the following lessons of relevance to the 'protection' of traditional knowledge:

- It is important that the public interest objective for intellectual property protection is paramount, i.e. to promote invention, and the application of inventions for the benefit of society at large.
- As a corollary, if the financial interests of the beneficiaries of protection are given primacy, then this is likely to detract from the wider societal benefits.
- In reality, the net financial benefits from protection are likely to be small. Inflated expectations of financial gain cause conflicts with those who wish to make use of knowledge or genetic resources.
- The introduction of protection tends to disrupt established processes for producing knowledge. It may lead to internal conflicts in institutions/communities and to reluctance to share knowledge internally or externally.
- The introduction of protection, or other restrictions such as those imposed through MTAs, tends to lead to greater restrictions on the use of basic materials and knowledge which are principally further inputs into the research process, rather than embryonic final products to benefit consumers.

Conclusions

The argument presented here is that we need to approach very carefully the extension of intellectual property rights into new fields, be it traditional knowledge or the world of scientific academia. The experience of introducing the intellectual property system into the academic world offers some indication of how this can transform, and also threaten to undermine, the traditional ways by which knowledge has developed. The system may provide incentives that might not otherwise exist for the development and application of technologies produced by academic scientists. But it can also distort research priorities, increase transaction costs, make some technologies less accessible to other researchers and undermine the fine balance that exists between cooperative and competitive behaviour in academia that contributes to the development of knowledge. The pursuit of material reward may displace the pursuit of knowledge as the primary motor of the system.

To be fair, the advocates of patenting on the lines of the Bayh–Dole Act were always clear that the purpose was to accelerate the transfer of technology from academia and its application for the benefit of the economy and society. Financial benefits that might accrue as a result were incidental to this purpose. That the participants in the system are often not clear about this distinction is a product of the way the intellectual property system uses private material rewards to stimulate creation or innovation. Thus the practice of applying patenting often tends to deviate from the theory, in particular because the private incentives on offer are only very loosely aligned with what would be required to fulfil the public policy objectives of IPRs. The instrument is a very blunt one.

In the case of traditional knowledge, it is evident that the argument for introducing intellectual property, for most advocates, is about equity and fair compensation for the use of knowledge or genetic resources that 'belong' to particular communities, and only a very distant second about incentives for continued innovation or conserving traditional knowledge that may be disappearing. The logic of Bayh–Dole, if not its practice, is reversed. Nor is it at all clear that an intellectual property instrument is an effective way of dealing with the decline of communities (and the knowledge associated with them), which has deep-seated economic and social causes.

Moreover, because communities that hold traditional knowledge tend to be poor – and vulnerable to wider economic, social and political forces – their traditional system of knowledge production, and way of life, is far more fragile than that of publicly-supported scientists in academia. Thus, the impact of introducing intellectual property, and the associated change that it induces in the way knowledge is produced, may have more serious negative impacts than those alleged in the case of Bayh–Dole on academia. In addition, to the extent that it places restrictions on, or adds to the cost of, traditional knowledge previously treated as being in the public domain, this will tend to impact negatively on follow-on invention and application of traditional knowledge for public benefit (analogous to the possible impact of patenting on upstream academic research, or on the cost of medicines).

We need to consider carefully the implications of introducing a new form of intellectual property to protect traditional knowledge and genetic resources. This is not to say that there are not good arguments for rewarding those who provide such knowledge with a share of the benefits that might arise from their commercial exploitation by others, but this is a separate issue (at least conceptually) from according intellectual property rights to the holders of such knowledge. Similarly, there are forceful arguments for modifying the existing patent system to avoid the problems associated with 'biopiracy', and to facilitate the sharing of benefits with the providers of knowledge or resources. Defensive measures, such as cataloguing traditional knowledge so it is accessible to patent examiners, also have an important part to play in preventing 'biopiracy'.

However, all these measures come with a cost including, as we have seen, national access regimes stimulated by the CBD. If the real problem for many communities is poverty and the disruption of their traditional way of life as a result of economic change, then measures that directly address those problems would be the most appropriate. This would avoid the deleterious, and possibly counterproductive, impact, for both local communities and the world at large, of introducing a new intellectual property regime for traditional knowledge.

Notes

1. The author, at the time of writing was seconded from the U.K. Department for International Development to the World Health Organization, as Secretary to the Commission on Intellectual Property Rights, Innovation and Public Health. The views expressed here are the author's own and do not necessarily reflect the views of any of these organisations.
2. We use the term 'traditional knowledge', rather than local or indigenous knowledge, because this is the common usage in international discussions on the use of intellectual property rights in this area, e.g. the World Intellectual Property Organisation's Intergovernmental Committee on Intellectual Property and Genetic Resources, Traditional Knowledge and Folklore (WIPO IGC). We use the term 'modern science' in a broadly Popperian sense.
3. For a discussion of the relationship between the two, from the perspective of modern science, see ICSU (2002).
4. This discussion does something of an injustice to the intricacies of concepts of ownership in local communities. It is argued that the concept of knowledge being in the 'public domain' is foreign to how such communities view their knowledge, even if it is widely distributed. Their rights in the knowledge are not extinguished by the act of publicising it, even when it is done with their informed consent. For a discussion of this, see Dutfield (1999).
5. The article also acknowledges less positive aspects arising from academic patenting.
6. In 1999, it is estimated that net research funding from licence income amounted to $149 million out of total R&D expenditure in U.S. academic institutions of $30 billion (Commission on Intellectual Property Rights 2002: 124).
7. These points are from a presentation: Nelson (2001).
8. In a footnote to this quotation the Review notes that 'MIT only earns around 3% of its revenues from licence income and fees'.
9. As of October 2004, there were only twelve national access regimes implementing the CBD, and two regional arrangements (International Chamber of Commerce 2004: 8).
10. See Massarani (2003). However, the government has subsequently responded by relaxing the rules for genuine scientific research (De Oliveira 2003).

References

Agres, T. 2003. 'Biodiversity Treaty Called Disastrous', *The Scientist*, 10 September 2003. www.biomedcentral.com/news/20030910/03

Commission on Intellectual Property Rights 2002. *Integrating Intellectual Property Rights and Development Policy*. London: Commission on Intellectual Property Rights. www.iprcommission.org

Convention on Biological Diversity 1992. www.biodiv.org/convention/articles.asp

Dasgupta, P. and P. David 1994. 'Towards a New Economics of Science', *Research Policy* 23: 487–521.

Department of Health and Human Services 2004. 'Best Practices for the Licensing of Genomic Inventions', *Federal Register* 69 (223), 19 November 2004. http://a257.g.akamaitech.net/7/257/2422/06jun20041800/edocket.access.gpo.gov/2004/pdf/04-25670.pdf

De Oliveira, W. 2003. 'Brazil Eases Restrictions on Biodiversity Researchers', *SciDevNet*, 12 November 2003. www.scidev.net

Dutfield, G. 1999. 'The Public and Private Domains: Intellectual Property Rights in Traditional Ecological Knowledge', *Electronic Journal of Intellectual Property Rights*, Oxford IP Research Centre Working Paper 03/99. http://www.oiprc.ox.ac.uk/EJWP0399.html

GRAIN 2003. 'The TRIPS Review at a Turning Point?' Source: www.grain.org/docs/trips-july-2003-en.pdf

ICSU 2002. 'Science and Traditional Knowledge Report from the ICSU Study Group on Science and Traditional Knowledge'. http://www.icsu.org/ Gestion/img/ICSU_DOC_DOWNLOAD/220_DD_FILE_Traitional_Kn owledge_report.pdf

International Chamber of Commerce (ICC) 2004. 'Access and Benefit-Sharing for Genetic Resources: Discussion Paper October 2004'. http://www. iccwbo.org/home/statements_rules/statements/2004/212-12E.pdf

Jaffe, A. 2000. 'The U.S. Patent System in Transition: Policy Innovation and the Innovation Process', *Research Policy* 29(4–5): 531–57.

Lambert, R. 2003. *Lambert Review of Business-University Collaboration: Summary of Consultation Responses and Emerging Issues Paper*. London: HM Treasury. http://www.hm-treasury.gov.uk/media//06729/lamberte-mergingissues_173.pdf

Massarani, L. 2003. 'Brazil's Biopiracy Laws "Are Stifling Research"', *SciDevNet* 21 July 2003. www.scidev.net.

Mowery, D., R. Nelson, B. Sampat and A. Ziedonis 2001. 'The Growth of Patenting and Licensing by U.S. Universities: an Assessment of the Effects of the Bayh-Dole Act of 1980', SIPA Working Paper 99–5. http://www.sipa.columbia.edu/RESEARCH/Paper/99–5.pdf

Nelsen, L. 1998. 'The Rise of Intellectual Property Protection in the American University', *Science*, 279(5356): 1460–61. http://www.sciencemag.org/ cgi/content/full/279/5356/1460

Nelson, R. 2001. 'The Contribution of American Research Universities to Technological Progress in Industry'. Handout at REITI Policy Symposium, 11 December 2001. http://www.rieti.go.jp/en/events/01121101/nelson.pdf

Rai, A. and R. Eisenberg 2003. 'Bayh-Dole Reform and the Progress of Biomedicine', *Law and Contemporary Problems* 66(1/2): 289–414. http://www.law.duke.edu/journals/lcp/downloads/lcp66dWinterSpring200 3p289.pdf

Rattray, G. 2002. 'The Enola Bean Patent Controversy Biopiracy, Novelty and Fish-and-Chips', *Duke Law and Technology Review*, Duke L. & Tech. Rev. 0008. http://www.law.duke.edu/journals/dltr/articles/PDF/2002DLTR0008.pdf

Revkin, A. 2002. 'Biologists Sought a Treaty; Now They Fault It', *New York Times*, 7 May 2002.

Royal Society 2003. *Keeping Science Open: The Effects of Intellectual Property Policy on the Conduct of Science.* London: The Royal Society. http://www.royalsoc.ac.uk/files/statfiles/document-221.pdf

WTO Council for TRIPS 2003. *Taking Forward the Review of Article 27.3(B) of the TRIPS Agreement,* Joint Communication from the African Group. IP/C/W/ 404. Geneva, 26 June 2003. http://docsonline.wto.org/ DDFDocuments/t/IP/C/W404.doc

11 Farmer Knowledge and Scientist Knowledge in Sustainable Agricultural Development: Ontology, Epistemology and Praxis

David A. Cleveland and Daniela Soleri

Introduction

What comprises local scientific knowledge of traditional or indigenous farmers (FK) and formal global scientific knowledge (SK)? How similar are they? What is 'sustainable' agriculture and what roles should FK and SK play in sustainable agricultural development? Who determines these roles and what effect does the assignment of roles have on the success of development projects? These are some of the questions that we have been asking ourselves and others during our years spent working with farmers and scientists in applied research and development in many locations around the world.

Conventional agriculture is widely acknowledged to be unsustainable, and more sustainable ways of producing food are advocated both for industrial and traditionally based agriculture (Matson et al. 1997; Tilman et al. 2002; Boody et al. 2005). However, sustainable agricultural development is a goal, based on values (see Sillitoe, this volume). It increasingly involves participation of both farmers and scientists, and thus requires an understanding of FK and SK. To respond more effectively to the needs of small-scale farmers in the Third World, we need to discuss openly the values underlying different definitions of sustainability to reach consensus on goals of agricultural development, and the empirical basis of definitions of FK and SK to understand their potential roles in meeting these goals.

FK and SK about the biophysical world are often defined deductively, based on assumptions that follow from the definition of sustainability used in a given programme or project (see Rhoades and Nazarea, this volume). This conflates two realms that are ontologically and epistemologically distinct: (1) sustainable agriculture, which can only be defined as a subjective, value-based goal, and cannot, therefore be objectively verified (although sustainability under a given definition can be evaluated using objective indicators), and (2) FK and SK about the biophysical world, which are not goals, but are concepts about the world which can be objectively verified independently of any definition of sustainable agriculture by comparing them to exogenous systems of rationality (both farmers' and scientists') – descriptions of reality agreed on by farmers and scientists – and by measuring their efficacy in meeting goals (e.g. of 'sustainability') when translated into practice. While some argue that FK and SK cannot be usefully distinguished because together they constitute different aspects of a 'hybrid' knowledge (see Dove et al., and Smith, this volume), we believe that even if FK and SK are part of a larger knowledge system, it is useful to compare them analytically. In this way both similarities and differences can be compared in terms of reference to a common ontological model (Soleri and Cleveland 2005), and differences in knowledge among farmers (Soleri et al. 2002) and among scientists (Cleveland 2001) can be examined. The practical importance of this conflation is that roles of farmers and scientists in development are often determined by the deductive definitions of FK and SK deriving from value-based goals, and not by an understanding of FK and SK based on empirical research. When the roles of farmers and scientists are based on untested assumptions about the nature of FK and SK, the probability of attaining the goal of sustainability, under any given definition, may be significantly reduced. This is especially important because of the growing interest in the potential of FK to make a contribution to agricultural development, both to increase the effectiveness of scientist and farmer research and practice, and to empower farmers.

We focus on small-scale, Third World farmers cultivating in marginal environments, using minimal external inputs, most of whom are poor – hereafter simply 'farmers'. Food production by these farmers is important to meet the growing demand for food (Narayanan and Gulati 2002), even with the expected increase in production in large-scale agriculture in more optimal environments (Heisey and Edmeades 1999). It is estimated that by 2025 three billion people will depend on small-scale Third World agriculture production (Falkenmark 1994: cited in Evans, 1998 #2325; Goklany 2002).

There has been much written about the way in which the concepts and use of local indigenous knowledge and FK held by different people and groups – e.g. NGOs (non-governmental organisations), environmentalists, government organisations, scientists, the media, and indigenous farmers themselves – depend on their social and political positions, especially the misunderstanding of FK by scientists and Westerners in general (e.g., Ellen et al. 2000; Sillitoe 2000; Haverkort et al. 2003). Building on this discussion, we present arguments below that the dominant views of FK and SK, are based, at least in part, on different definitions of sustainable agriculture, and each implies different roles for farmers, natural scientists and social scientists involved in sustainable agricultural development.

Sustainable Agricultural Development: Ontology, Epistemology and Knowledge

Defining and Measuring Sustainable Agriculture

Many agricultural development policies and projects, from the World Bank and the FAO (Food and Agriculture Organization) to the smallest grassroots farmer organisation, are today labelled as 'sustainable'. All these definitions share a common desire for agriculture to 'develop' in a way that is sustainable in the basic sense of not self-destructing over the short term, but beyond this definitions can diverge radically. One inherent difficulty is that sustainability is a goal, a teleological concept that cannot be measured until a subjective, value-based definition has been agreed on (Costanza 2001). This is often not sufficiently recognised, resulting in discussions that do not deal with the basis of disagreement. All of the key components in any definition of sustainable agriculture, including the spatial and temporal boundaries of the system, and which are the 'good' aspects to be kept and the 'bad' to be eliminated, are subjective judgements (c.f. Thompson 1995).

Confusion about these basic ontological and epistemological aspects of sustainable agriculture abound. For example, an examination of sustainable agriculture in Zimbabwe asserted that a definition of sustainable agriculture was impossible, but then used objective indicators to measure sustainability (Campbell et al. 1997). However, the indicators chosen, including crop yield and soil organic matter, imply a definition of sustainability, and the authors' conclusion that indicators of sustainability are always inadequate may result in part from the failure to define sustainability in the first place. In addition, their claim that objective biophysical criteria for sustainable agriculture are impossible to define because they are overwhelmed by social and political changes external to them, could be addressed by including social and economic as well as environmental criteria in the definition.

Sustainable agriculture, and sustainable development in general, are often conceived of as having three main components: economic, environmental and social (e.g., Goodland 1995; Costanza 2001). However, one of these components is often emphasised over the other two, as illustrated in Table 11.1. Each of these emphases is based on contrasting underlying assumptions about the human carrying capacity of the Earth and the components of human impact (population size, consumption levels, and the technology used to produce what is consumed) (Daily and Ehrlich 1992). In addition, there are important assumptions about some of the major variables that affect the components of human impact: natural resources, human nature and markets (Costanza 2001).

Table 11.1 Definitions of sustainable agriculture with different emphases, and the assumptions they are based on.

Sustainability components	Environmental emphasis	Social emphasis	Economic emphasis
		Definitions of sustainability	
Sustainability components	Conserving natural environment (ecocentric), to provide resources for people (anthropocentric). Focus on natural capital.	Social justice, empowerment of indigenous peoples, women, minorities. Focus on social (moral) and human capital; fair distribution.	Continuous growth of economy to provide wealth for future generations. Focus on economic (human-made) capital; allocative efficiency.
		Assumptions	
Human carrying capacity	Surpassed (or will be soon)	Best dealt with later	No limit (in near future), can be increased for a very long time.
Human population size	Needs to be controlled	Not the major problem	Not a problem
Consumption	Depends on population size, must be reduced	Inequity is main problem	Must increase
Technology	Must be more efficient, less impact on environment	Must be more equity of access	Can be improved to increase human carrying capacity
Natural resources	Complementary; finite physical (source and sink) limits to supply.	Unequal distribution major problem	Substitutable; demand and technology drive supply
Human nature	Concern for environment	Concern for others	Concern for self
Markets	Can't value natural resources; destroy environment. Must define limits ecologically. Low discount rate on future value.	Can't value social good; destroy community. Need redistribution to address inequity. Low discount rate on future value.	Translate self-interest into social good. Trickle down to address inequity. High discount rate on future value.

When agreement on a subjective definition of sustainable agriculture is reached among a group of people (explicitly or implicitly), then indicators of sustainability can be generated, and the degree to which a given component of a specific agricultural system, including practices and knowledge (FK and SK), is sustainable can be objectively assessed. However, the choice of indicators will also necessarily be influenced by values, meaning that while measuring sustainability can be done more objectively than defining sustainability, measuring sustainability can never be completely objective, just as no knowledge of objective reality can ever be completely objective.

Knowledge and Sustainable Agriculture

The belief that indigenous knowledge is critical for sustainability has been spreading for more than two decades. An important milestone in this movement was the incorporation of this concept in the 1992 Convention on Biological Diversity. For example, Article 8(j) calls for signatories to 'respect, preserve and maintain knowledge, innovations and practices of indigenous and local communities embodying traditional lifestyles relevant for the conservation and sustainable use of biological diversity' and to 'encourage the equitable sharing of benefits' arising from the use of same.

Sustainable agriculture is central to sustainable development, and there is much interest in the potential for synergy between SK and FK, one of the central components of the debate over global vs. local knowledge. Farmers whose well-being and way of life is threatened by modern technology opportunistically make use of possibilities offered by modern technology to improve their situation (Cleveland 1998) – or they will no longer be able to remain farmers (see Dove et al., this volume, on hybrid knowledge). They may define their agriculture in ways that include industrial agriculture technologies, in part because it serves their larger goal of maintaining their physical and cultural identity. For example, Zuni indigenous farmers have learned how to use global positioning system (GPS) technology to map their family farm fields, and this has become a powerful force in resolving land disputes that have impeded the revitalisation of their farming system (Cleveland et al. 1995). Farmers' advocates, including many local and international NGOs, are also promoting the inclusion of FK in agricultural development (e.g. Haverkort et al. 2003).

Scientists are also interested in using FK to increase the sustainability of agriculture. Many have suggested that some local lessons can be generalised to the global scale, for example in management of common pool resources such as irrigation water, over which global competition is rapidly increasing (Ostrom et al. 1999). Matson et al. (1997: 508) advocate 'the development of more ecologically designed agricultural systems that reintegrate features of traditional agricultural knowledge and add new ecological knowledge'.

The explicit inclusion of both FK and SK in agricultural development is usually initiated by scientists or development professionals, probably because they

are the ones in power. The result is participatory research and development programmes, with the implication that farmers are participating in scientists'
applied research and development (Soleri et al. 2002). Multilevel or multistage
classifications of participation are common in this kind of research, and tend to
emphasise the degree of social and institutional participation of farmers and scientists (e.g., Biggs 1989). Often, the roles of farmers, natural scientists and social
scientists appear to be determined by implicit or explicit assumptions about what
constitutes FK and SK that derive deductively from the definitions of sustainable
agriculture employed, not from empirical understanding of the nature of FK and
SK in the contexts involved.

Farmer and Scientist Knowledge in Sustainable Agriculture Development

The most prominent views of the nature of FK and SK and their roles in development reflect those of the wider discussion of local or indigenous knowledge
and modern, global science. These views tend to be based on unexamined, often
value-based assumptions about the nature of these knowledges. Most can be classified into three broad categories based on definition of FK (Table 11.2): the economically rational farmer, the socioculturally rational farmer, and the ecologically rational farmer. Blaikie et al. (1997) propose a classification of local knowledge in natural resource development in which their 'classical' corresponds
roughly to our 'economically irrational' view of farmers (not included in our
Table 11.2, but see brief discussion below), 'neoliberal' to our 'economically
rational', and 'neopopulist' to our 'socioculturally rational' and 'ecologically
rational'. A fourth view, held by a minority, is one which we term the complex
farmer. In the following sections we briefly describe these views of FK and the
correlated views of SK, suggesting how each is based at least in part on a different definition of sustainable agriculture, and how each implies different roles for
farmers, natural scientists and social scientists involved in sustainable agricultural development (Table 11.2).

The Economically Rational Farmer

Until after the Second World War a view of farmers as economically irrational
dominated Western ideas of Third World agricultural development. As research
on farmers increased, in part in an effort to understand their 'irrational' response
to development, this assumption was replaced by the view that farmers are economically rational, but limited environmentally socially, and economically.
Today farmers' 'behaviour may often seem irrational to Western economists who
have little comprehension of the precarious nature of subsistence living and the
importance of avoiding risks' (Todaro 1994: 282). As a widely-used textbook on
economic development states:

Table 11.2 Definitions of sustainable agriculture, farmer knowledge (FK), and scientist knowledge (SK) in relationship to the roles of actors in agricultural development.

View of farmer based on sustainable agriculture definitions	Definitions			Roles		
	Sustainable agriculture	FK	SK	Farmers	Natural scientists	Social scientists
1. Economically rational	Economic emphasis; monolithic model of unilineal progress towards modern agriculture; absorb FK into SK, local farmers into global economy	Economically rational but 'primitive', theoretically limited	More rational, theoretically more powerful	Integrate into modern agriculture (or eliminate)	Develop improved technologies	Convince farmers to adopt modern technologies; train and direct farmers to participate
2. Socioculturally rational	Sociocultural emphasis; farmers gain local autonomy to renaturalise nature	Natural, organic, contextual, a 'skill', atheoretical	Reductionist, destructive of nature, theoretically ignorant of FK	Continue indigenous farming	None	Help farmers in struggle against hegemony of SK
3. Ecologically rational	Ecological emphasis; environmental FK and/or SK can increase sustainability of agriculture	Ecologically rational, descriptive and discriminatory, codifiable, may be generalisable	(a) Ecologically less accurate, more destructive, or (b) Ecologically more accurate, less destructive	Develop more ecologically rational farming	(a) Learn ecological principles from farmers, or (b) Provide resources and expertise to improve farmer practice	Catalogue and disseminate FK
4. Complex	Holistic emphasis; new, syncretic forms of knowledge, collaboration in practice	Complex: intuitive, empirical, theoretical; more informal	Complex: intuitive, empirical, theoretical; more formal	Investigate and integrate scientists' approach	Investigate and integrate farmers' approach	Investigate farmer and natural scientist knowledge, facilitate communication

in spite of the relative backwardness of production technologies ... the fact remains that given the static nature of the peasants' environment, the uncertainties that surround them, the need to meet minimum survival levels of output, and the rigid social institutions into which they are locked, most peasants behave in an economically rational manner when confronted with alternative opportunities. (Todaro 1994: 305)

The rational farmer viewpoint that dominates mainstream agricultural development today is based on a unilineal theory of development, the highest level of which is the modern industrial state which enjoys mass consumption and other 'blessings and choices opened up by the march of compound interest' (Rostow 1971: 6). Agriculture is specialised and totally commercial, 'no different in concept or operation from large industrial enterprises' (Todaro 1994: 310).

Definition of sustainable agriculture. The definition of sustainable agriculture that characterises this view of farmers emphasises economics, and modern, technologically complex, high-input agriculture. For example, a fundamental assumption often made by genetically engineered (GE) crop advocates is that Third World agriculture is 'primitive' and that the major goal of agricultural development is ultimately to replace it with modern industrial agriculture, including genetically engineered crop varieties, incorporating farmers into the global seed system dominated by private companies (DeVries and Toenniessen 2001; Conway 2003).

Economic growth is a key component of sustainable agriculture, and the emphasis is on increasing the slope of total factor productivity, or output through time (Lynam and Herdt 1992), often based on modern crop varieties that are highly responsive to increased inputs such as irrigation and manufactured fertilisers and pesticides (Cleveland 2001). For example, a report on a development project in Senegal concluded that, 'If farmers are given better access to information, rice technologies, inputs and decision making, rice production on irrigated land in West Africa may leap forward rapidly as potential production gains are still large' (Haefele et al. 2002).

Population growth is taken as exogenous, and since sustainable agriculture must feed a growing population, it equates with 'sustainable growth' (Lynam and Herdt 1992: 211). The time period for measuring sustainability must be short enough to make a projection with low probability of error, i.e. less than twenty years, and the spatial scale must be limited, i.e. the farming system in a specific location, since higher organisational levels cannot be adequately defined. Environmental sustainability is subordinate to economic growth, e.g. the adoption of agro-ecological technologies is seen as dependent on whether farmers view them as increasing profit or welfare, and therefore will only be useful if they 'complement the continued use of inputs in the intensification of farming systems' (Lynam and Herdt 1992: 215).

Definition of farmer and scientist knowledge. This view of the rational farmer developed within economics in the 1960s in reaction to the view that farmers are 'irrational' or 'primitive'. The economists Boserup (1965) and Schultz (1964) published important books citing evidence to support their claims that farmers are capable of responding in economically rational ways to forces generated by the market place and population pressure.

Even as definitions of farmers' economic rationality continue to be refined, the underlying assumption remains that farmers attempt to maximise their individual utilities, making decisions in the same way that any business person would, if they have the same information and opportunities (Hardaker et al. 1997). Thus, seemingly irrational behaviour can be understood as the result of the constraint of 'partial engagement in … markets which are often imperfect or incomplete' (Ellis 1993: 13). Agricultural modernisation and development is cast in terms of improving markets, prices, technology or education to remove constraints on farmers' potentially economically rational behaviour that will lead them down the path to modern agriculture (Todaro 1994). It also focuses on replacing inferior FK with superior SK, as for example in a participatory plant-breeding project in Mexico, which attempted to teach farmers basic maize reproductive biology and selection techniques, assuming that in-field plant selection will be more efficient than the traditional method of selection of ears post-harvest (CIMMYT 2000). Often, the only FK considered worth researching is knowledge received from outsiders, as in an irrigation project in Senegal that documented farmers' knowledge of production practices recommended by the government irrigation and extension authority (Haefele et al. 2002).

Roles in agricultural development. Farmers' roles are passive – they are expected to give up their primitive ideas and methods and to adopt modern farming methods, or to get out of farming altogether. However, farmers may be considered to be 'only dimly aware of the potential benefits of improved germplasm and crop management practices', and lack the education and skills needed to manage modern crop varieties 'properly' (Aquino 1998: 249). Outsiders facilitate the replacement or modernisation of small-scale farmers, including replacement of their crop varieties with modern ones (Srivastava and Jaffee 1993). This is the dominant theme, for example, among both private companies and international agencies promoting GE crop varieties for Third World farmers, such as the Rockefeller Foundation policy for GE crops, which is similar to its policy for the Green Revolution (Conway 2003). Small-scale farms in the Third World are considered inferior and transitory, but requiring modernisation in the short run to keep people from migrating to cities (Hazell 2004). They need time to become educated enough to leave 'unproductive' farming behind.

The Socioculturally Rational Farmer

In part a response to the economic rationality viewpoint, the 'socioculturally rational farmer' perspective rejects the assumption that SK is always superior to FK, and that unilineal, market-driven agriculture development can be sustainable. Instead it emphasises the social and political relations believed to be implicit in conventional agricultural development, and proposes alternatives based on what proponents perceive to be the social and cultural perspectives of the farmers themselves.

Definition of sustainable agriculture. Definitions of sustainable agriculture emphasising social aspects and equity are often associated with a view of farmer knowledge as natural, organic, contextual, and skill rather than theory. These definitions often include the assumption that modern agriculture is inherently unsustainable, and indigenous agriculture inherently sustainable. Like the economically rational view, and in contrast to the ecologically rational view, this view does not emphasise environmental limits and the need to limit human impact.

Definition of farmer and scientist knowledge. This viewpoint proposes that FK and SK are – often fundamentally – different, and that FK is more sustainable because it is more 'natural', 'organic' and 'holistic' than SK, and farmers may be considered to be inherently conservators of their environment and their crop production resources (e.g., Escobar 1999). The difference in knowledge is based on an 'enormous epistemological difference' between 'peasants' whose thought is 'inherently holistic and dependent on identifying things in terms of their relationships to larger wholes' and that of the modern Western world which is 'inherently atomistic and reductionist, defining identity in terms of the thing itself and not in relation to the context of which the thing is part'(Taussig 1977: 150). Much of the discussion of knowledge systems and development is cast in modern vs. postmodern/premodern terms, emphasising FK and SK as fundamentally different, which often means that there can be no constructive combination of the two, and we are forced to choose between them.

There is often an assumption that knowledge is more socially than environmentally constructed, and the unique localness of FK is emphasised. For example, 'All traditional knowledge systems use different paradigms, which manifest themselves in the knowledge of everyday life …' (Haverkort et al. 2003: 36). Fairhead and Leach's study of West African farmers' management of their forest-savannah vegetation and soils concludes that they 'enrich' the landscape by converting savannah into 'forest islands', while providing themselves a relatively good living, and that local 'specificities' are most important (1996). They see this as a stark contrast to the 'reading' of development professionals who are allied with existing power structures, and see the landscape as degraded into patchy savannah from a pre-existing pristine forest by destructive land use. Fairhead and Leach link this 'reading' with functionalist equilibrium 'cultural ecology' models that they say are based on inadequate and outdated ecological theory and embedded in a Western science epistemology that decouples 'natural and social phenomena'.

One prominent approach that supports the socioculturally rational farmer view is the relativist, utilitarian tradition of ethnobiology or folk biology, proposing that local knowledge is unique and depends on the goals, theories and beliefs of the local people (Medin and Atran 1999: 6). As Berlin notes, the utilitarian tradition is often dominated by economic concerns, or descriptions of uses, and this continues to be a strong tradition in economic botany and zoology (Berlin 1992).

While the utilitarian approach in ethnobiology often emphasises the usefulness of SK as an aid in describing FK, a more extreme version of the socioculturally rational view sees attempts to explain farmers' knowledge and practice in scientific terms as impeding true appreciation of their knowledge (see Selener 1997: 175–76), with the implication that there is no ontological basis for comparison. Therefore, FK is relegated to a black box that can only be described by outside ethnographers, and it is not considered valid to investigate the relationship of FK to external reality or to SK, or the way it is generated. 'Culture' becomes an entity that is 'assimilated in something of the same way a body warms to the sun' (Medin and Atran 1999: 6). For example, Ingold rejects knowledge as economic rationality (embodied for him in evolutionary psychology) and knowledge as ecological rationality (embodied for him in evolutionary ecology) and advocates knowledge as acquired through performance, or 'enskillment', which seems to imply that farmers acquire knowledge through direct experience and contact with nature, rather than explicitly (e.g., Ingold 1996). Richards applies these assumptions to farmers' cropping patterns and sees each farmer's crop mixture as a 'completed performance' which can only be interpreted by 'reconstructing the sequence of events in time', because he declares that it is 'not the outcome of a prior body of "indigenous technical knowledge"' and 'much of it should be judged and valued not by the standards of scientific analysis, but as self-help therapy' (Richards 1993: 67, 70).

Roles in agricultural development. The farmers' role in sustainable agriculture in this view is often to continue their practices based on their traditional or indigenous knowledge. The 'proper' role for outsiders is empowerment of local people, and they 'must choose between being facilitators for local autonomy … by brokering the preservation and application of knowledge systems that contribute to rehumanization and re-naturalization of nature …, or be agents of hegemonic "progress"' (Purcell 1998: 267). Yet some who see universal, reductionist SK and local, holistic indigenous knowledge as fundamentally different, do see them sharing analogous processes and practices at a deeper level, and advocate debate between the two (e.g., Watson-Verran and Turnbull 1995).

The Ecologically Rational Farmer

The ecologically rational farmer view sees FK as ecologically rational, emphasising its descriptive and discriminatory value. However, while FK is often considered codifiable, and, therefore, to some extent generalisable, it is not generally regarded as theoretical.

Definition of sustainable agriculture. The ecologically rational farmer view gives the definition of sustainable agriculture an environmental emphasis. Like the socioculturally rational farmer viewpoint, it rejects the assumption that unilineal, market-driven agriculture development can be sustainable. This view, in contrast to the previous ones, often emphasises natural limits to growth, and the need to limit human impact through greater understanding of ecological principles to improve management.

Definition of farmer and scientist knowledge. In contrast to the socioculturally rational farmer viewpoint, and similar to the economic rationality viewpoint, the ecologically rational viewpoint sees no fundamental differences in FK and SK, because the 'subject matter' of agriculture is common to them both, and 'may be of much more importance than are the social and cultural contexts' (Sumberg and Okali 1997: 150). Sometimes this view explicitly contrasts farmers' ecological rationality with ecological irrationality engendered by conventional 'rational choice' economics (Chambers et al. 1989). This view emphasises farmers' accurate and sustainable ecological knowledge of their environments. There are two main variants of the ecologically rational farmer viewpoint. The first assumes that FK is usually superior to SK because farmers have an intimate knowledge of their environments, and seek out and emphasise areas of empirical and epistemological overlap between indigenous farmers and modern agricultural science. This has been critiqued as a 'populist' assumption that ignores the role of experience and intuition in local knowledge (Scoones and Thompson 1993). The second assumes that SK is usually superior to FK because of the greater explanatory power of modern science. It often sees farmers as 'barefoot scientists', whose knowledge needs to be vetted in comparison to the more accurate SK. (See (a) and (b) in row 3 in Table 11.2.)

The first version often sees SK as inferior because it is inextricably associated with capitalist economic assumptions (see Sillitoe, this volume). For example, a study of Zapotec farmers in Mexico suggests they are scientists because they 'hypothesize, they model problems, they experiment, they measure results, and they distribute knowledge ...', even though they 'typically proceed from markedly different premises – that is, from a different conceptual basis', one that is 'culturally incommensurable with those predominating in industrialized societies' (González 2001: 3) (see Smith, this volume, on incommensurability).

The second version suggests to some that the similarity of local and scientific knowledge is due to cognitive human universals as well as predictable regularities in the natural world (Boster 1996). The intellectualist or comparativist tradition in ethnobiology takes this view. It sees categories as recognised rather than culturally constructed because nature herself comprises an independent organised pattern, and there are universals in human cognition, resulting in cross-cultural similarities in the ways in which humans conceive biological organisms (Medin and Atran 1999: 8). It is sometimes assumed that SK can serve as the ontological comparator, or the 'metalanguage in terms of which the folk system can be understood' (Berlin 1992: 201). Some behaviour is seen as influenced by group dynamics at a level at which farmers may not be cognizant, for example in the management of large-scale irrigation systems (Lansing et al. 1998).

Roles in agricultural development. In the first version of this view, the role of outsiders is to understand the extent that FK is compatible with SK, and to support and learn from farmers. For example, participatory plant-breeding projects may focus on improvements based on farmers' crop varieties, with the implicit or explicit assumption that these varieties and FK are locally adapted (Soleri et al. 2002).

In the second version, the role of outsiders is to figure out inadequacies in FK and to remedy them with the application of SK. For example, participatory plant-breeding projects may focus on improvements based on plant breeders' modern crop varieties, with the implicit or explicit assumption that these varieties and SK can be locally adapted and will, therefore, result in more ecologically sustainable development (Soleri et al. 2002). Often outsiders use SK to evaluate FK, as in a study of soil erosion in Burkina Faso, which concluded that farmers have a good knowledge of wind erosion processes, but not of water erosion processes, and are willing to apply new techniques to control erosion, but the main constraints to apply these measures are insufficient knowledge and lack of labour (Visser et al. 2003).

The Complex Farmer

This is the viewpoint we believe is most consistent with the available information on FK and SK. The main difference between the complex farmer viewpoint and the three just described is that it emphasises distinguishing as much as possible value-based from empirically based knowledge, and an inductive approach to understanding FK and SK. This opens up the possibility that FK and SK can be both similar and different, and that either one can be 'better' than the other, depending on the situation and definition of sustainable agriculture.

Definition of sustainable agriculture. Definitions of sustainable agriculture are holistic, including sociocultural, environmental and economic sustainability, and explicitly discuss the empirical and value-based assumptions underpinning the definition.

Definition of farmer and scientist knowledge. An important goal is to understand similarities and differences between local and global scientific knowledges in a practical way. The interest is not only in the extent to which local and global scientific knowledges are similar or different, but also whether one is *better* than the other, not in a metaphysical sense, but in an empirical and practical one – that is, for any specific situation we need to ask 'Which knowledge, SK, FK or a combination of SK and FK, produces the most sustainable agriculture given our definition?' (c.f. Medin and Atran 1999: 12). As Sillitoe states, 'The objective is not to assess the veracity of local ideas against scientific ones, both are relative, but to enrich our overall understanding of environmental interactions within cultural contexts' (Sillitoe 1996: 11). The outcomes can be measured in terms of both farmers' and scientists' goals.

Based on our review of existing research, the general hypothesis on which this view is based is that both FK and SK (1) include knowledge about how reality *ought to be*, based on individual values and goals, and knowledge about how reality *is*, based on observations of external social and biophysical reality; (2) contain knowledge about how reality is that is *localised* and empirical because based on unique local contexts, and *generalised* and theoretical because based on widespread patterns (e.g. due to biological evolution); and (3) are *conscious* and *unconscious*. For example, rather than assuming that FK is tacit and embodied in contrast to SK which is conscious and explicit, the complex farmer view interprets research on FK and SK as showing that both are in important ways both tacit and explicit (Sillitoe, this volume, Chapter 1). As Scoones and Thompson stated it, FK and SK 'are both general and specific, theoretical and practical. Both are value-laden, context-specific and influenced by social relations of power' (1994: 29).

Similarities between SK and FK result from the common biophysical environment experienced by both scientists and farmers, and the biological similarities in physical perception and cognitive function shared by all humans. Differences between SK and FK result from the many unique characteristics of farmers' situations compared with those of scientists, including different growing environments, crop genotypes, cultural values and social organisation. A number of examples of this approach exist, although not necessarily identified explicitly in the terms we use here. They demonstrate a high degree of variability in and between FK and SK, underlain by consistent patterns, providing the basis for complementarity and collaboration.

In northwest Syria, FK and SK of soils and land use potential was compared and found to be complementary, with FK more local and emphasising sociocultural variables, whereas SK was more general and emphasised biophysical variables (Cools et al. 2003). Research on farmers' weather prediction and their use of scientific meteorological information in Burkina Faso found that farmers operate in multiple cognitive frameworks, and are interested in receiving scientific information because they perceive local forecasting methods as becoming less reliable as a result of increasing climate variability. However, there are significant differences between scientific and local forecasts: the former predict total rainfall quantity at a regional scale, whereas the latter stress rainfall duration and distribution and are more attuned to crop–weather interactions and stress the relationship between knowledge and social responsibility (Roncoli et al. 2002). Malawian farmers' taxonomy of cassava varieties based on plant morphology distinguished varieties with no morphological differences between them perceptible to scientists, but whose distinctness was supported by molecular analyses for cyanogenic glucoside levels, and by fine-grained genetic analysis (Mkumbira et al. 2003). It appears that their extensive experience and observations have resulted in FK of cassava plant morphology being more extensive and capable of discriminating at a more subtle level than SK.

Research also shows FK, like SK, to include theory. In reviewing the results of research on subsistence of rainforest peoples, including his own with the Nuaulu of Seram, eastern Indonesia, in terms of their knowledge of nature, Ellen concludes that observations of 'particular instances' (substantive knowledge of many individual species) leads inductively to 'knowledge of general principles', and in knowledge transmission these 'overarching deductive models of how the natural world works are privileged over accumulated inductive knowledge'(Ellen 1999: 106). These models function at a macro-scale as a 'folk synecology', for example in connecting observations at the species level with forest structure and dynamics (Ellen 1999: 107). Wola farmers of New Guinea are aware of the geomorphological forces that both destroy and renew their soil resources, and can use their knowledge of processes to explain future aspects of soil formation (Sillitoe 1996: 135–36). However, unlike scientists, they do not appear to be aware of processes on a geological time scale. There are also similarities and differences between the taxonomic classification of animals by the Wola of New Guinea and Western science (Sillitoe 2003: 62–71). While Sillitoe states that 'identification depends on a fundamentally different approach to classification' (2003: 71), his data suggest that there are also fundamentally similar aspects.

In our recent research we have used scenarios based on basic biological principles to elicit FK (Soleri and Cleveland 2005) of genotype-by-environment interaction, heritability and genetically engineered crops varieties (e.g. Soleri et al. 2002; Soleri et al. 2005). While no truly neutral ontological comparator exists, we chose the basic biological model of genotype–environment relationships, which is universally accepted by biologists, including plant breeders, although they disagree among themselves about its *interpretation* at higher levels of generalisation, for example whether selection in optimal or marginal environments leads to genotypes that are better adapted to marginal environments (Cleveland 2001; Ceccarelli and Grando 2002). This variation in scientists' interpretations suggests that, when farmers do in fact think in terms of the basic model, it can be a valuable comparator, facilitating the consideration of FK and SK on equal grounds.

Overall our results are consistent with the complex farmer view – there are patterns in FK across different crops and countries, and between FK and SK that support the hypothesis that empirical and theoretical FK and SK consistently reflect similar patterns and relationships in reality; but there are also differences among farmers, among scientists, and between FK and SK, and these can often be explained in terms of differences in reality, e.g., in crop varieties, or in cultural values.

Roles in agricultural development. The challenge for scientists and development professionals changes in the complex farmer viewpoint from emphasising a deductive definition of knowledge, to emphasising inductive empirical research to understand the complexities that determine knowledge and practice in a particular situation, in order to promote sustainable agriculture as defined. The role of farmers will then depend on the extent of similarities and differences in FK and SK and the particular problem being addressed (Cleveland and Soleri 2002).

For example, goals in our research are to learn (1) how farmers understand the basic biological model of relationships between plant genotypes and growing environments that determine plant phenotypes including the results of seed selection; (2) how this understanding affects farmers' practices and expectations; (3) how FK of the basic biological model is similar to or different from SK of this model, and (4) how to contribute to collaboration between scientists and farmers to find ways to improve the results of plant breeding in farmers' own terms.

Can a Better Empirical Understanding of Knowledges Enhance Sustainable Agriculture?

If achieving a more sustainable agriculture requires that we develop new, syncretic forms of knowledge and practice, ones based on both modern agricultural SK, as well as the traditionally based FK, then we need to understand more about the nature and functioning of both FK and SK.

Real-world decisions are always based on incomplete knowledge, i.e., they are risky decisions (Hardaker et al. 1997). But decisions have to be made – and we should try to improve the likelihood that we are making the 'right' decisions by distinguishing between goals about the way things ought to be, in the form of definitions of sustainable agriculture, and empirical understanding of the way things are, in the form of increased understanding of FK and SK. We propose that distinguishing the different ontological natures of sustainability and knowledge could increase our ability to debate explicitly stated definitions of sustainable agriculture in terms of values, and the empirical nature of SK and FK in terms of data and hypothesis testing. In this way SK and FK are more likely to be successful tools for achieving whatever vision of sustainable agriculture we may agree on.

Acknowledgements

We thank the many farmers and scientists we have worked with in Cuba, Egypt, Ghana, Guatemala, Mexico, Mali, Nepal, Pakistan, Syria and the United States for sharing their knowledge both about the way things are, and their hopes for the way things should be. We thank the UCSB Faculty Senate, the U.S. National Science Foundation (SES-9977996, DEB-0409984), and the Wallace Genetic Foundation for recent support of research.

References

Aquino, P. 1998. 'Mexico', in *Maize Seed Industries in Developing Countries* (ed.) M.L. Morris. Boulder, CO and Mexico: Lynne Rienner and CIMMYT.

Berlin, B. 1992. *Ethnobiological Classification: Principles of Categorization of Plants and Animals in Traditional Societies.* Princeton, NJ: Princeton University Press.

Biggs, S.D. 1989. *Resource-poor Farmer Participation in Research: a Synthesis of Experiences from Nine National Agricultural Research Systems.* (OFCOR Comparative Study Paper. The Hague: International Service for National Agricultural Research (ISNAR).

Blaikie, P., K. Brown, M. Stocking, L. Tang, P. Dixon and P. Sillitoe 1997. 'Knowledge in Action: Local Knowledge as a Development Resource and Barriers to its Incorporation in Natural Resource Research and Development', *Agricultural Systems* 55: 217–37.

Boody, G., B. Vondracek, D.A. Andow, M. Krinke, J. Westra, J. Zimmerman and P. Welle 2005. 'Multifunctional Agriculture in the United States', *Bioscience* 55: 27–38.

Boserup, E. 1965. *The Conditions of Agricultural Growth.* Chicago, IL: Aldine.

Boster, J. 1996. 'Human Cognition as a Product and Agent of Evolution', in *Redefining Nature: Ecology, Culture and Domestication,* (eds), R. Ellen and K. Fukui. Oxford: Berg.

Campbell, B.M., P. Bradley and S.E. Carter 1997. 'Sustainability and Peasant Farming Systems: Observations from Zimbabwe', *Agriculture and Human Values* 14: 159–68.

Ceccarelli, S. and S. Grando 2002. 'Plant Breeding with Farmers Requires Testing the Assumptions of Conventional Plant Breeding: Lessons from the ICARDA Barley Program', in *Farmers, Scientists and Plant Breeding: Integrating Knowledge and Practice,* (eds), D.A. Cleveland and D. Soleri. Wallingford, Oxon, U.K.: CAB International.

Chambers, R., A. Pacey and L.-A. Thrupp. 1989. *Farmer First: Farmer Innovation and Agricultural Research.* London: Intermediate Technology Publications.

CIMMYT (Centro Internacional de Mejoramiento de Maiz y Trigo) 2000. The Oaxaca project: farmers conserving maize diversity in the farmers' fields. Project home page: http://www.cimmyt.org/Research/economics/oaxaca/overview/ov_pa.htm

Cleveland, D.A. 1998. 'Balancing on a Planet: toward an Agricultural Anthropology for the Twenty-first Century', *Human Ecology* 26: 323–40.

———— 2001. 'Is Plant Breeding Science Objective Truth or Social Construction? The Case of Yield Stability', *Agriculture and Human Values* 18(3): 251–70.

Cleveland, D.A., F.J. Bowannie, D. Eriacho, A. Laahty and E.P. Perramond 1995. 'Zuni Farming and United Stated Government Policy: the Politics of Cultural and Biological Diversity', *Agriculture and Human Values* 12: 2–18.

Cleveland, D.A. and D. Soleri 2002. 'Introduction: Farmers, Scientists and Plant Breeding: Knowledge, Practice, and the Possibilities for Collaboration', in *Farmers, Scientists and Plant Breeding: Integrating Knowledge and Practice*, ed, D.A. Cleveland and D. Soleri. Wallingford, Oxon, U.K.: CAB International.

Conway, G. 2003. *From the Green Revolution to the Biotechnology Revolution: Food for Poor People in the 21st Century.* New York: Rockefeller Foundation.

Cools, N., E. De Pauw and J. Deckers 2003. 'Towards an Integration of Conventional Land Evaluation Methods and Farmers' Soil Suitability Assessment: a Case Study in Northwestern Syria', *Agriculture Ecosystems and Environment* 95: 327–42.

Costanza, R. 2001. 'Visions, Values, Valuation, and the Need for Ecological Economics', *BioScience* 51: 459–68.

Daily, G.C. and P.R. Ehrlich 1992. 'Population, Sustainability, and Earth's Carrying Capacity', *BioScience* 42: 761–70.

DeVries, J. and G. Toenniessen 2001. *Securing the Harvest: Biotechnology, Breeding and Seed Systems for African Crops.* Wallingford, Oxon, U.K.: CAB International.

Ellen, R. 1999. 'Models of Subsistence and Ethnobiological Knowledge: between Extraction and Cultivation in Southeast Asia', in *Folkbiology*, (eds), D.L. Medin and S. Atran. Cambridge, MA: MIT Press.

Ellen, R., P. Parkes and A. Bicker (eds) 2000. *Indigenous Environmental Knowledge and Its Transformations: Critical Anthropological Perspectives.* Amsterdam: Harwood Academic Publishers.

Ellis, F. 1993. *Peasant Economics: Farm Households and Agrarian Development.* Cambridge: Cambridge University Press.

Escobar, A. 1999. 'After Nature: Steps to an Antiessentialist Political Ecology', *Current Anthropology* 40: 1–30.

Evans, L.T. 1998. *Crop Evolution, Adaption and Yield.* Cambridge: Cambridge University Press.

Fairhead, J. and M. Leach. 1996. *Misreading the African Landscape: Society and Ecology in a Forest-Savanna Mosaic.* Cambridge: Cambridge University Press.

Falkenmark, M. 1994. 'Landscape as Life-support Provider: Water-related Limitations', in *Population, the Complex Reality*, (ed.) F. Graham-Smith. London: The Royal Society.

Goklany, I.M. 2002. 'From Precautionary Principle to Risk-risk Analysis', *Nature Biotechnology* 20: 1075.

González, R.J. 2001. *Zapotec Science: Farming and Food in the Northern Sierra of Oaxaca.* Austin: University of Texas Press.

Goodland, R. 1995. 'The Concept of Environmental Sustainability', *Annual Review of Ecology and Systematics* 26: 1–24.

Haefele, S.M., M.C.S. Wopereis and C. Donovan 2002. 'Farmers' Perceptions, Practices and Performance in a Sahelian Irrigated Rice Scheme', *Experimental Agriculture* 38: 197–210.

Hardaker, J.B., R.B.M. Huirne and J.R. Anderson 1997. *Coping with Risk in Agriculture*. Wallingford, Oxon. U.K.: CAB International.

Haverkort, B., K. van 't Hooft and W. Hiemstra 2003. *Ancient Roots, New Shoots*. Leusden, The Netherlands and London: ETC/Compass and Zed Books.

Hazell, P. 2004. 'Last Chance for the Small Farm?' *IFPRI Forum* October 2004, 7–8.

Heisey, P.W. and G.O. Edmeades 1999. 'Part 1. Maize Production in Drought-stressed Environments: Technical Options and Research Resource Allocation', in *World Maize Facts and Trends* 1997/98 (ed.) CIMMYT. Mexico, D.F.: CIMMYT.

Ingold, T. 1996. 'Hunting and Gathering as Ways of Perceiving the Environment', in *Redefining Nature: Ecology, Culture and Domestication*, (eds), R. Ellen and K. Fukui, pp. 117–56. Oxford: Berg.

Lansing, J.S., J.N. Kremer and B.B. Smuts 1998. 'System-dependent Selection, Ecological Feedback and the Emergence of Functional Structure in Ecosystems', *Journal of Theoretical Biology* 192: 377–91.

Lynam, J.K. and R.W. Herdt 1992. 'Sense and Sustainability: Sustainability as an Objective in International Agricultural Research', in *Diversity, Farmer Knowledge, and Sustainability*, (eds), J.L. Moock and R.E. Rhoades. Ithaca, NY: Cornell University Press.

Matson, P.A., W.J. Parton, A.G. Power and M.J. Swift 1997. 'Agricultural Intensification and Ecosystem Properties', *Science* 277: 504–9.

Medin, D.L. and S. Atran 1999. 'Introduction', in *Folkbiology*, (eds), D.L. Medin and S. Atran. Cambridge, MA: MIT Press.

Mkumbira, J., L. Chiwona-Karltun, U. Lagercrantz, N.M. Mahungu, J. Saka, A. Mhone, M. Bokanga, L. Brimer, U. Gullberg and H. Rosling 2003. 'Classification of Cassava into "Bitter" and "Cool" in Malawi: From Farmers' Perception to Characterisation by Molecular Markers', *Euphytica* 132: 7–22.

Narayanan, S. and A. Gulati. 2002. *Globalization and the Smallholders: a Review of Issues, Approaches, and Implications*. Discussion Paper. Washington, D.C.: International Food Policy Research Institute (IFPRI).

Ostrom, E., J. Burger, C.B. Field, R.B. Norgaard and D. Policansky 1999. 'Revisiting the Commons: Local Lessons, Global Challenges', *Science* 284: 278–82.

Purcell, T.W. 1998. 'Indigenous Knowledge and Applied Anthropology: Questions of Definition and Direction', *Human Organization* 57: 258–72.

Richards, P. 1993. 'Cultivation: Knowledge or Performance?', in *An Anthropological Critique of Development: the Growth of Ignorance,* (ed.) M. Hobart. London: Routledge.

Roncoli, C., K. Ingram and P. Kirshen 2002. 'Reading the Rains: Local Knowledge and Rainfall Forecasting in Burkina Faso', *Society and Natural Resources* 15: 409–27.

Rostow, W.W. 1971. *The Stages of Economic Growth.* Cambridge: Cambridge University Press.

Schultz, T.W. 1964. *Transforming Traditional Agriculture.* New Haven, CT: Yale University Press.

Scoones, I. and J. Thompson 1993. *Challenging the Populist Perspective: Rural People's Knowledge, Agricultural Research and Extension Practice.* Discussion Paper. Brighton, U.K.: Institute of Development Studies, University of Sussex.

———— 1994. Knowledge, 'Power and Agriculture – towards a Theoretical Understanding', in *Beyond Farmer First,* (eds), I. Scoones and J. Thompson. London: Intermediate Technology Publications.

Selener, D. 1997. *Participatory Action Research and Social Change.* Ithaca, NY: The Cornell Participatory Action Research Network, Cornell University.

Sillitoe, P. 1996. *A Place against Time: Land and Environment in the Papua New Guinea Highlands.* Amsterdam: Harwood Academic Publishers.

———— (ed.) 2000. *Indigenous Knowledge Development in Bangladesh: Present and Future.* London: Intermediate Technology Publications.

———— 2003. *Managing Animals in New Guinea: Preying the Game in the Highlands.* London: Routledge.

Soleri, D. and D.A. Cleveland 2005. 'Scenarios as a Tool for Eliciting and Understanding Farmers' Biological Knowledge', *Field Methods* 17: 283–301.

Soleri, D., D.A. Cleveland, F. Aragón Cuevas, H. Ríos Labrada, M.R. Fuentes Lopez and S.H. Sweeney 2005. 'Understanding the Potential Impact of Transgenic Crops in Traditional Agriculture: Maize Farmers' Perspectives in Cuba, Guatemala and Mexico', Environment Biosafety Research 4: 141–66.

Soleri, D., D.A. Cleveland, S.E. Smith, S. Ceccarelli, S. Grando, R.B. Rana, D. Rijal and H. Ríos Labrada 2002. 'Understanding Farmers' Knowledge as the Basis for Collaboration with Plant Breeders: Methodological Development and Examples from Ongoing Research in Mexico, Syria, Cuba, and Nepal', in *Farmers, Scientists and Plant Breeding: Integrating Knowledge and Practice,* ed, D.A. Cleveland and D. Soleri. Wallingford, Oxon. U.K.: CAB International.

Srivastava, J.P. and S. Jaffee 1993. *Best Practices for Moving Seed Technology: New Approaches to Doing Business.* World Bank Technical Paper. Washington, D.C.: The World Bank.

Sumberg, J. and C. Okali 1997. *Farmers' Experiments: Creating Local Knowledge.* Boulder, CO: Lynne Riener.

Taussig, M. 1977. 'The Genesis of Capitalism amongst a South American Peasantry: Devil's Labor and the Baptism of Money', *Comparative Studies in Society and History* 19: 130–55.

Thompson, P.B. 1995. *The Spirit of the Soil: Agriculture and Environmental Ethics.* London and New York: Routledge.

Tilman, D., K.G. Cassman, P.A. Matson, R. Naylor and S. Polasky 2002. 'Agricultural Sustainability and Intensive Production Practices', *Nature* 418: 671–77.

Todaro, M.P. 1994. *Economic Development.* New York: Longman.

Visser, S.M., J.K. Leenders and M. Leeuwis 2003. 'Farmers' Perceptions of Erosion by Wind and Water in Northern Burkina Faso', *Land Degradation and Development* 14: 123–32.

Watson-Verran, H. and D. Turnbull 1995. 'Science and Other Indigenous Knowledge Systems', in *Handbook of Science and Technology Studies*, ed, S. Jasanoff, G.E. Markle, J.C. Petersen and T. Pinch. Thousand Oaks, CA: Sage Press.

12 Forgotten Futures: Scientific Models vs. Local Visions of Land Use Change

Robert E. Rhoades and Virginia Nazarea

The future is a central theme in the global agendas of sustainability science and sustainable development. After decades of short-term research and planning based on three- to five-year project budgeting plans or annual cropping cycles, scientists under *Agenda 21: Programme of Action for Sustainable Development* were mandated to shed light on sustainability's central question of 'preserving for future generations the same opportunities available to our generation' (World Commission on Environment and Development 1987: 8). One limitation of conventional development, and the underlying science, has been the discounting of long-term negative impacts on the environment of short-run human goals, behaviour, and policies. In contrast, sustainable development's future horizon for planning was projected to be minimally twenty to thirty years in the future, with an emphasis on 'action and social learning' involving the full participation of local communities. Under the sustainability vision, science and development could no longer proceed in their comfortable spatiotemporal frameworks but now had to deal head-on with future spaces inhabited by future generations.

A question rarely asked in sustainability academic and applied circles, however, is: how does this futuristic component of sustainability play out in local communities, especially indigenous ones, where concepts of time, space and local values about the future may differ? Do local people think of time as a linear trajectory with a past which leads to the present and continues to the future? Is there common ground between the people's vision and values about the future and those of scientists? How, for example, do local communities look upon land use change (LUC) modelling, the favourite tool of sustainability science, which predicts or projects future conditions based on trend analysis over several past decades? Do futuristic GIS (Geographic Information Systems) maps and charts of land use strike a familiar chord in the local social and cultural context, or are they strange images of Western science which are politely viewed in community gatherings but internally rejected? If, indeed, a goal of sustainability science is the involvement of multiple voices in place-based research, planning and policy, the question of how time – especially future time – is understood, recognised and negotiated becomes a central research and operational issue.

Through a comparative case study in two distinct ethnoecological regions in Northwest Ecuador, this chapter explores differences and similarities of scientists' and local people's conception of a sustainable future. The scientific viewpoint is represented by land use change (LUC) analysis, often considered the cornerstone of many natural resource and agriculture sustainability projects. To get at the deep seated principles which govern local people's views of time, their landscape and the future, however, innovative ethnoecological elicitation techniques based on local culture were utilised. The two perspectives – scientific and local – are then contrasted through a future-visioning methodology to seek a common platform for debate and discussion on sustainability questions.

People and Landscapes of Northwestern Ecuador

Northwestern Ecuador offers an ideal set of human and environmental conditions for pursuing global science and community-based development and conservation. The conservation centrepiece of the region is the 204,000 hectare Cotacachi-Cayapas Ecological Reserve, a prime wilderness area extending from the alpine pastures of Mount Cotacachi (4939 metres above sea level, masl) in the western cordillera, to the western humid lowland forest not far from the Pacific Coast (under 500 masl). From summit to sea, the reserve and adjacent area form a transect covering eleven tropical biozones. Within this highland-lowland ecological system, the variety of endemic plant and animal species renders the region among the world's highest priorities for conservation. The area contains two of the world's 'hotspots' characterised by an extraordinarily high species number per unit area (e.g., 11 species per 1,000 m^2; 15 bird species per km^2 (Alarón 2001). Inside the reserve are hundreds of critical watersheds supporting endangered species of mammals and birds, including the spectacle bear, jaguar, ocelot, mountain tapirs, monkeys, plate-billed mountain toucan, and the Andean condor (Rhoades 2001).

The diverse landscape has a corresponding human lifescape of rich cultural diversity. Within and near the reserve's buffer zone live four distinct ethnic groups pursuing different livelihood and land use strategies. Towards the east in the high Andean zone are Quichua-speaking Indian communities of Cotacachi, while located south of the reserve are *mestizo* frontier colonists (*colonos*). In the western lowlands towards the Pacific Ocean in the Chocó-Andean forest are Chachi Indians and Afroecuadorians. These ethnic groups are economically poor, possess little to no land, and have limited access to external resources. The population growth rate of 2.1 percent is one of the highest in Latin America.

In prevailing development and conservation discourse, human settlements and the natural environment of Northwest Ecuador are portrayed as being on a collision course (Sierra 1996). This ecological 'paradise', according to the international pleas to save it, is threatened by smallholder farmers who are migrating toward the Cotacachi-Cayapas Ecological Reserve in search of new land. While the reserve's interior remains largely unaffected by local human impact, the primary forests in the adjacent buffer zones are rapidly disappearing. Conservationists fear that these marginal settlers, especially along the southern forest frontier zones and the long-established Andean communities to the east, will destroy through ill-informed farming practices the native forest or Andean alpine grass ecosystems (*paramo*). This degradation scenario, employed as a justification for outside funding and intervention, makes a clear causal linkage between population dynamics and pressure on the land (Rhoades 2001). Subsistence farmers, lacking adequate productive land or other options for return on their labour, are portrayed as involuntarily forced up the slopes or deeper into the forests in search of virgin farmland.[1] Clearing the forest as they settle, they leave a degraded landscape of fragmented forest patches characterised by high rates of soil erosion, lower fertility, reduced biodiversity, and other negative environmental consequences. Farmers presumably abandon their old fields to search for more fertile land towards and within the reserve. According to conservation rhetoric, this unsustainable cycle of land degradation repeats itself with each migration wave, leading to a downward spiral for the whole region unless interventions such as ecological corridors, protected areas, environmental education, and sustainable agriculture are instituted.

Given the widely accepted discourse about rampant human forest destruction in Northwest Ecuador, the region is swarming with international conservation and development projects aimed at slowing migration into the primary forests (Rhoades and Stallings 2001). Sustainable development in the region is a multi-million dollar industry for NGOs (non-governmental organizations), government agencies, private ecological reserves and international development/conservation projects. In addition to the Cotacachi-Cayapas Ecological Reserve, the region supports at least six privately owned nature reserves which are aligned with international conservation projects. Large-scale integrated conservation and development projects (ICDPs) funded by USAID, World Bank and other multilateral agencies are everywhere (Rhoades and Stallings 2001). One large Global Environmental Fund (GEF) project aims to construct a one million hectare Chocó-Andean Corridor which would connect private reserves to the main Cotacachi-Cayapas Ecological Reserve and beyond to Colombia (Chocó-Andean Project 2000).

Our sustainability research programme, Sustainable Agriculture and Natural Resource Management (acronym: SANREM)[2] selected research sites in two distinct critical buffer zones where the presumed threat by smallholders to the protected area was proclaimed to be greatest (Figure 12.1). Nanegal to the south is a colonisation hillside parish (c. 1,000 masl) inhabited mainly by *mestizos* of mixed European and Amer-Indian descent. Andean Cotacachi (2,500–4,000 masl), situated on the eastern side of the reserve, is populated by 20,000 indigenous Quechua-speaking people living in forty communities (*comunas*). These two cases, located within the same buffer zone of a major reserve, represent contrasting sustainability challenges due to strikingly different agricultural practices, social organisation, technologies, local knowledge systems and cosmologies (see also Heckler, this volume, on differences within indigenous knowledges).

Nanegal is a typical South American zone of mature colonisation created by in-migration from other rural areas. Over the past half century, migrants have cleared the forest and established sugar cane fields, areas of mixed cropping and pastures for cattle (Rhoades 2001). The population maintains a strong link to the market based on sugar cane, distilled alcohol and cattle. Nanegal is characterised by isolated homesteads, small service towns, and limited government services. In contrast, the Quechua-speaking indigenous people of Cotacachi have lived for centuries in inter-Andean communities around the base of Cotacachi volcano mountain. Large *haciendas* historically dominated the landscape until land reform in the 1960s and 1970s when indigenous people, who had worked as virtual serfs (called *huasipungeros*) on these large estates, received titles to small parcels of land equivalent to what they had worked prior to land reform. Given the decline in agricultural productivity and farm prices, most young people today engage in circular labour migration to Quito and nearby towns and marketing of artisan crafts. Through the 1980s and 1990s, the indigenous communities have increasingly organised themselves politically and socially to press for continued land reform, bilingual education and, most recently, development (Rhoades 2006). Cotacacheños have a strong cosmological and ritual connection to the landscape not found in Nanegal, including attachment to particular communities, respect of sacred sites, and a ritual calendar aimed at maintaining ethnic identity and building group solidarity.

Land Use Change Modelling: Scientific and Local Frameworks

Land use change (LUC) modelling has been central to the new suite of decision support tools tested and promoted by sustainable science researchers (Lambin et al., 2003). Successful modelling requires good quantitative data of a biophysical, economic or spatial nature and seeks to reduce complexity to a simpler order. Data-intensive, robust LUC models typically present 'if-then' hypotheses and scenarios or trade-offs between development and environmental impacts (e.g., if you build a road, then you lose x percent of forest over x number of years). The advantage of LUC analysis is that it compresses and simplifies long-term land changes into mathematical expressions and predictions so that users can perceive complex processes and extended time horizons which are cognitively difficult, if not impossible, for humans. LUC analysis helps scientists and planners understand system drivers and derive land use change rules, which predict – for example – how a road will affect forest cover or how growth of a population centre is related to type of crop land.

The LUC modelling approach is conceptually straightforward (Stewart 2001). Land use maps are prepared from aerial photographs or satellite images, ground-truthing of vegetation cover, and other available studies that describe in some detail land use on at least two points in time (e.g. today and twenty years ago). The more points available, the better for analysis, but the number used is typically determined by level of image resolution, reliability and availability of original data sources. The hectares assigned to each land use category are tabulated for each year and the change in each is calculated for different years. The transition rates describing changes observed between the two time points (at a minimum) are then utilised to create a third set of category sizes (types of land use) that represent the land use in the future, say 2030. Supplemental information on roadways, trails, waterways and community boundaries allow the modeller to analyse human impacts on land use. With this information digitised and placed in GIS, a distanced analysis can be made illustrating how much forest is lost or pasture created as a result of various human interventions. This analysis of 'roads/land use change' (or other drivers such as population, markets, migration) can be used to extrapolate projections a generation hence. Other calculations are typically made in the LUC analysis process, including development of detailed diagrams showing the rates of change as percent of original hectares converted to other uses over the years as well as land use combinations acting as 'sinks' that will inevitably dominate the landscape to the exclusion of other land uses.

While the scientific value of LUC modelling is recognised, it has some drawbacks for community-based research. Its predictive abilities are often overblown, clients rarely understand the underlying assumptions, issues of interest to science can lead to neglect of other 'unmodelled' processes, and users rarely scrutinize the quality of the data (Honachefsky 2000; Rhoades 2001). Modelling often takes place independently of local decision makers who presumably use the model. Shabman (1995) puts it this way: 'as model building has become more complex and costly, models have become the domain of experts who are expected to develop models independent of decision makers and then transfer the model's results to those decision makers'. Although rarely admitted by modellers, scientific modelling is commonly generated for priority dissemination to other scientists in journals, books and at professional meetings. Policy makers are another presumed audience but it is debatable how and when policy makers use models vis-à-vis political pressures in decision making (King and Kraemer, 1992). 'Modelling for modelling's sake' and sheer intellectual excitement and professional advancement seem to drive most academic modelling. The rush to modelling in sustainable development projects comes from a desire to be more 'scientific' through prediction, simulations and expert systems. Proponents of LUC have also been able to garner a significant amount of development funds, given the heavy data input demands, technological support and personnel needed to develop and apply models. Modellers often have the expectation that 'the model' becomes the 'core' of research and all data collection should 'serve the model'. A well-conceived model, backed by mesmerising multimedia presentation, can easily place viewers under its spell and leave the impression that the model reflects 'reality'. As Sillitoe (this volume) points out for science in general, models are also cultural constructs. As such, they emphasise aspects of interest to scientists and neglect others in the 'reality' they are trying to represent (e.g., focus on forest change). The external pressure to use such models in sustainable development is high, although critical questions about their effectiveness have not been satisfactorily answered by proponents of modelling.

In our view, the true test of a model's efficacy rests on whether scientists can translate their findings into a format where local people can recognise themselves and their landscape in the past and future projections. Human subjects of models, especially rural populations, are rarely given a chance to understand or test them on their terms or in the environment where the model was generated. If the model did not work, inform, or stimulate good debate by the local beneficiaries, what use would it be if extrapolated to faraway places? This commonsense requirement of local verification is rarely applied to test a LUC model's potential for decision making. Our team's goal was to design a participatory approach where both scientists and local people compare and hopefully combine their respective knowledge and expertise to explore alternative pathways to the future through a careful consideration of understanding how past and present behaviour impacts the environment and lifestyles (DeWalt 1994).

Mapping the Landscape: Searching for Culturally Relevant Images

One favourite method for comparing scientific and local knowledge in sustainable development is mapping (Poole 1995; Peluso 1995). For example, comparison of scientific soil maps with farmer soil maps provide useful information on differences as well as commonalities in soil classification and use (Talawar and Rhoades 1998; Sillitoe et al. 2004). Ideally, maps function to compress complex physical reality into easily understood, simplified images, although the ability to read maps varies by culture and formal education. In sustainability projects, LUC analyses are nearly always presented to donors, colleagues and local people in attractive GIS time-series colour maps which specify how land use categories have changed in the past and how they are projected to change in the future.

Are GIS maps the most appropriate way to communicate land use findings at the local level and, if not, are there alternative forms of communication? LUC scientists devote enormous energy and resources to production of their maps in various dimensions and scales (e.g., two-dimensional, digital terrain models, visual simulations). Few resources, however, are devoted to asking whether or not such digital-age maps and images are culturally relevant decision support tools for local people. To address this gap in sustainability mapping in our research sites, we asked local people to interpret seventeen kinds and scales of maps and images of their landscapes. Formal maps and landscape images (e.g., cartographic, aerial, satellite, soil maps and photos) were shown to different local groups (age and gender) to test their comprehension of the images. At the same time, we elicited landscape information through informal mapping (community-based and individual maps) produced on the spot with colour pencils and paper. In addition, participatory 3-D models of the landscape at 1:10,000 scale were constructed and used in interaction with local informants (Rhoades and Moates 2003).

Most available maps for our research areas, whether formal productions or local drawings, were largely static representations of land use situations at the time of map production. One exception was the participatory mapping exercise wherein people drew their landscape 'in the past', today and 'a generation in the future'. But these free-drawn maps were not geo-referenced and thus impossible to overlay or easily compare with scientific maps. Local maps by informants did illustrate salient features of importance to local people and gave an idea about desired future conditions. We experimented with many variants, ranging from providing some geo-referenced points to an outline of the landscape extracted from panoramic photographs and asking people to 'draw their landscapes' at various times. We discovered that when we assigned a few geo-referenced orientation points (e.g., church or mountain peak), people were reluctant to draw at all, declaring that the points confused them. The images produced by researchers which elicited the most interest and which could be geo-referenced and manipulated were the 3-D physical models and 180° panoramic photos of well-known local landscapes. While the digital panoramic photo obviously had no indigenous roots, the 3-D model (Spanish: maqueta) has been used in the Andes as far back as the Inca Empire to plan paths, buildings, canals and roads for settlement (Hyslop 1990: 116; Denevan 2001: 5). By asking people upfront what they did and did not understand about maps or images, we gained insights as to how we could effectively move from scientific land use modelling to culturally anchored visions. However, moving from mapping to participatory, people-led future visioning required a new approach for contrasting science and local knowledge.

Future Visioning by Local People and Scientists

Scientific visioning has become a major activity in planning and development (World Future Society 2002). Visioning and scenario building comes in many guises under different labels and is used in scales ranging from the local to the global. Examples include 'envisioning' for world society (Costanza 1997), desired future conditions for forest management in the northeast U.S. (Twery et al.1998), visioning for local communities in Oregon (Ames, 1989), exploratory land use studies for rural areas of the European Union (van Ittersum et al., 1998), and anticipatory visioning for forest users in tropical ecoregions (Wollenberg et al. 2001). Although proponents of future visioning and scenario modelling argue that the approach can be used anywhere and with anyone, virtually all reported case studies are from the U.S., Canada, Europe, Australia or New Zealand. We addressed this gap by developing and testing a methodology among rural peasants and indigenous groups where Western assumptions about time, space and values for the future may not apply.

Thinking about the future, whether called prediction, projection, envisioning or scenario building, is not only a favourite pastime of scientific specialists but is a cultural universal. All known societies care about and plan for the future of their children and subsequent generations (Textor 1999). Aristotle stressed the importance of each generation not to constrain the fulfilment of future generations (cited in Gerlagh and Papyrakis 2003).

Although there is no 'ethnography of the future' compared to the large literature on anthropology of time, sense of place and culture memory (e.g., Gell 1992; Basso 1996; Nazarea 1998, 2005), indigenous people have imagined and even acted upon creating better futures. Revitalisation movements such as Melanesian cargo cults, millenarian sects and the Native American ghost dance are ways in which suppressed indigenous groups have followed – often with disastrous results – prophets who promised a new world (Wallace 1970). Today, indigenous people's post-*Agenda 21* declarations at international gatherings stress the importance of envisioning a better world based on indigenous cosmology and values. Two widely quoted 'visioning' statements in the literature came from the Iroquois and Dene. The Iroquois law states that no council decision should be made unless it considers consequences in seven generations. The Iroquois constitution states:

> Look and listen for the welfare of the whole people and have always in view not only the present but also the coming generations, even those whose faces are yet beneath the surface of the ground – the unborn of the future nation. (Iroquois Tribe *c.*1475: 5)

The 'river of time' concept of the Dene Nation (Canada) evokes the metaphor of a flowing river:

> We take our strength and our wisdom and our ways from the flow and direction that has been established for us by our ancestors we never knew, ancestors of a thousand years ago. Their wisdom flows through us to our children and our grandchildren to generations we will never know. We will live out our lives as we must and we will die in peace because we will never know that our people and this river will flow on after us. (T'selele 1977: 14)

Whether science-driven or created by local people, visioning or future scenarios are simply, in the words of Wollenberg et al. (2001: 331), 'stories of what might be' and are used 'to evaluate what to do now based on different possible futures'. Whether arrived at through scientific extrapolation from current trends, simulation modelling using changing exogenous variables, or simply speculating about the future based on cultural values, scenarios reflect a group's 'mental construct of ways of looking at the future'. It is difficult, however, to get local visions in contexts where science is not culturally or contextually translated.

People are often reluctant to predict or even talk about the future in concrete terms, especially the rural poor. People accustomed to a lack of control over their lives may prefer to acknowledge the power of or defer to fate, luck, or God's will rather than to make predictions. There may be a need to develop a willingness among the audience to face uncertainty and to understand the forces driving it. (Wollenberg et al. 2001: 338)

Based on our ethnographic research in Nanegal and Cotacachi we conclude that this presumed 'reluctance' comes mainly from researchers being unwilling to recognise that their modelling results are social constructs of their own making. Talking or visioning their own compartmentalised view of the future may well lead to silent groups of locals who feel uncomfortable or intimidated by the symbols and language of science. In the Andes, where communication is largely visual and verbal – not written or mathematical – people have a long tradition of story telling, including stories about future events. As Sillitoe (this volume) notes for the New Guinea system of counting, the challenge for our team was to discover how future time works within the within the Andean cultural framework and contrast it with the scientific projections. Without an available methodology in our project to address this communication gap, we decided to initiate parallel lines of enquiry which would converge in a scientific-local dialogue about the future. Scientific modellers began modelling long-term change of the landscape, while team anthropologists started with ethnoecological methods to elicit the deep-seated principles that structure a local people's assessment and management of their environment. The following steps were taken in our emerging 'future-visioning methodology'.

Step 1. Model the Scientific View of Landscape Change. Analyse and explain scientifically the past, present and future land use change over forty years (e.g., 1960–2000) by studying aerial photos, remote sensing, and secondary reports of target landscapes. Link LUC to 'human drivers' (population, roads, markets, etc.) to describe the dominant historical process that impacts the landscape (forests, farmland, pasture, etc.) through time. Project these dominant trends and processes into the future through simulation to derive land use rules and scenarios. Generate GIS maps of past, present and future land use based on the simulation model.

Step 2. Elicit the Indigenous View of Landscape Change. Using ethnoecological methods, we researched local understanding of landscape changes over the same time period of the LUC model. In addition to key informant interviewing, participatory free-hand, time-series mapping, and community history workshops, we used two techniques to focus on the future. One was what we call 'cultural in-visioning' (a form of story completion based on oral traditions and model-derived rules of LUC). The second was a more focused drawing of past and future (desired) landscapes using panoramic perspectives with only horizon contour lines drawn; this blank perspective map was based on a well-known, easily visible landscape viewpoint recognisable to all community members. These two exercises provide a comparison of the scientists' scientific-derived vision (predictive statements) with local visions (cultural interpretations) to arrive at an understanding of differences in assumptions, values, beliefs, perceptions, and time and space horizons.

Step 3. Contrast Scientific and Local 'Futures'. A comparative dialogue takes place when the LUC modelling predictions of Step 1 are presented to the community in a locally understandable form. This involves transforming the scientific LUC patterns or predictions into a readily understood panoramic view of a locally recognised landscape, namely the conversion of the written, mathematical and GIS maps from the model into a locally recognisable oral and visual image. This reinterpreted scientific landscape, in turn, is compared to the landscape 'envisioned' by local people according to their own values (Step 2 above).

Understanding Scientific Scenarios through Folktales

Our first attempt to compare scientific and cultural landscape visions/scenarios took place among the *mestizo* campesinos (peasants) of Nanegal. To construct a model of LUC in this area, we first carried out a 24-year land use study from 1966–1990, based on aerial photographs and remote sensing images. These data were analysed along with information on roadways, waterways and community boundaries as human drivers in land management. 'Transition rates' of observed changes between 1966 and 1990 were used to project a future scientific scenario to 2030. It was assumed existing areas would grow or decline as dictated by the change in category of land use allocation, modified by the human drivers.

Team modellers assumed that roads provide access to agricultural resources and pastures (Stewart 2001). The model revealed a clear relationship between the geographic layout of the road and trail system and the spatial pattern of forest cover. This conclusion led to the hypothesis that road building facilitates shrinking of forested areas. Also, changes in land use would be expected in areas near population centres. Areas furthest from roads and communities would be less subject to change. This analysis enabled our modellers to derive general principles of land use change that can be used to build future landscape scenarios. In general, analysis of the data confirmed that land use follows the pattern of converting nearby forest into cropland and pasture. The model projects a decline in the amount and quality of forested areas in Nanegal. As colonisation proceeds, the process of deforestation will continue until only a few forest remnants remain on steep slopes. Further analysis shows that this relationship holds not because roads are built in already deforested areas but because road building is associated with the disappearance of the forest.

Team modellers recognised that road analysis is only one of many factors that may be used to identify significant trends in land use change and to create scientifically plausible future scenarios. The construction of a road is a culturally and economically significant event since it provides critical access to remote pasture, cropland and timber resources, as well as increased economic ties with neighbouring regions. Moreover, roads are directly associated with the loss of forest cover for all communities. Recently established communities show greatest loss near roads while older communities show increasing loss further from town centres. In broad terms, the future landscape of Nanegal, according to our model, will see increased conversion from one agricultural use to another and creation of *chaparral* (brush) as a result of abandonment of croplands and pastures. Reforestation may continue in pockets but it seems to have run its course. Although the model outputs are more statistically detailed than presented in this chapter, the 'derived land use change rules' for Nanegal can be summarised: (1) Forest distant from town becomes pasture; (2) forest close to town becomes mixed pasture/cropland; (3) inaccessible pasture distant from town will be abandoned to succession vegetation; (4) accessible pasture will become cropland.

But to what extent are the model's projections and derived LUC rules, especially forest conversion, a concern for local residents? Are they willing to change their management practices and forego development goals or their values to safeguard or restore the environmental integrity of the region? The modeller's focus on land use change, especially loss of native or primary forest, is justified in terms of international conservation concerns for this region, but are these priorities held by the people of Nanegal for their landscape and their future?

To answer these questions, we turned to ethnoecological methods. Unlike scientific modelling, which addresses biophysical impacts of human–environmental interactions, ethnoecological methods aim to understand the cultural, moral and aesthetic values that shape the way local people perceive and manage natural resources. We refer to our approach as 'cultural in-visioning' to stress the place-centre, micro-level dimensions of local projection. To elicit conscious and subconscious attitudes and aspirations concerning land use changes, we designed culturally relevant story completion tests using elements of local folktales and narrative traditions. A story completion test is a projective method used in cognitive anthropology and clinical psychology to 'bring out' the respondent's thoughts, motivations, and aspirations – conscious and subconscious sentiments, attitudes, and perceptions that he or she may find difficult to express under more direct questioning (Nazarea et al. 1998). Initially, we collected as many folktales (*cuentos*) as we could from people. Reading through the collected stories, we noted some common elements that animated the various plots. One was about a widow who seemingly illustrated an interesting tension between contradictory impulses, benevolence/generosity and self-righteousness/greed. Second was the opposing pair of good brother/bad brother. A third was the kind king or president who, unfortunately, was in the throes of imminent death and therefore creating great anxiety among his subjects. Finally, there was the 'guesser' whose 'clairvoyance' largely depended on previously planted crops. Although this region is inhabited primarily by *mestizos*, these folktale characters and motifs have clear Andean origins, some of which can be traced back to pre-Inca times.

To parallel the scientific modellers' time frame and rules, we constructed two story-lines that project thirty years into the past and thirty years into the future and encouraged informants to talk about changes in land use categories (e.g. forest, pasture, cropland, etc.). We wanted people to think about the past thirty years (to test out the model's historical trends) and also to project themselves into the future and identify overriding concerns, priorities in land use, future aspirations, urgently needed changes, and communication problems. Informal interviews were conducted with fifteen male and fifteen female Nanegal residents of different ages and socioeconomic standing. A decision was made to conduct the interviews and story-telling sessions in private to minimise the influence of others in the community. The modified folktales were read to the local respondents but the completion of the story and its questions was left entirely up to them. The responses were lively, engaging and informative about past and future landscapes. Any reluctance to talk about the past or future was minimised by the folktale context. Two examples of modified stories with gaps for respondents to complete were as follows:

Story 1: *The widow and the two brothers*
The brothers meet a needy widow who is begging. One offers her some food, the other insults her. The kind brother, who had been suffering from a limp, is healed. The other loses his senses and gets lost. He returned to the village after thirty years. He noticed the road and the other changes: He saw that the forest had _____ ; He saw that the pasture land had _____ ; He saw that the sugar cane had _____ ; He saw that the people had

_____ .

Story 2: *The Wise President and the Infallible Guesser*
A much beloved president is growing older and people begin to worry about what will happen. They consult a village sage who has a reputation of accurately predicting the future. But rather than giving them an answer, the sage decides to learn what is on people's mind and what concerns them thirty years into the future. This is what he finds out: The people are most concerned about _____ ; They work hard because _____ ; If something could be changed for the better, they would _____ ; They feared the new president would not understand _____ .

The first story completion test (widow and two brothers) corresponded to the thirty year historical time span on which the scientific modelling trend analysis exercise was based. The responses agree with the modelling results, roads and trails figure prominently in people's perceptions of a changing landscape, and corroborate the derived land use change principles. But, unlike scientists and environmentalists, most respondents perceived the conversion of forest to cropland and pasture as positive. The forest was not considered a resource to be preserved but one that must be exploited and managed for the benefit of the community. They stated that lack of attention and care, rather than a conservation motive, allows the land to revert back to forest. Labour is seen by locals as a more important determinant than proximity in driving forest conversion, livelihood priorities determining its allocation. One typical response to 'He saw that after thirty years the forest had …', was:

> … (the forest) disappeared and the land is more populated and the village is more developed. The people are kinder, they share more and with everybody. They are more caring. If there are sick people, they had meetings and collected money for someone to treat the patient. People share … hence, the village developed more.

In response to the question about what most concerns people in thirty years' time, we find the same concerns with community building, livelihood issues and work, not with forest preservation. While our LUC analysis, showing the eventual disappearance (if present trends continue) of all but remnants of forested area, was a cause for alarm among environmentalists, the story completion test revealed that local people have a different view. One typical response was: 'we are worried about work and problems with the economy and how to progress and develop for our children's future', while another person stated 'People are worried about how to get better and to work to improve the community', and another 'People worry most about improving their lives, especially in relation to our community. Like, for example, school and medical centre. We need to unite and collaborate.'

In all story completion responses about the future not a single informant was concerned about the forest disappearance. From the mestizo colonists' point of view their very raison d'être is to clear the forest and transform the landscape into a settled human one. They migrated to Nanegal for that purpose, not conservation. The stories emphasise that if cultivated parcels or pasture are allowed to revert back to brush (*chaparral*) or forest then the owner is lazy and negligent. This local conclusion is just the opposite of the global conservation agenda of scientists and environmentalists, who are concerned about the disappearance of the forest. Global green programmes are not local agendas. Future scenario stories by Nanegal families show that their main concern is finding work or increasing the productivity and profitability of local enterprises. They also look favourably upon viable income-generating alternatives, such as mining, floriculture and poultry production, even if they might carry negative environmental impacts. Work is seen as the key to progress, which is understood to involve better schools, health facilities and government services.

What we have seen from the results of the story completion test is that both modellers and local people see clear connections between the human condition and different land use and understand that specific outcomes depend on a host of factors. Modellers, however, tend to concentrate on factors that are easier to quantify, like distance from a road, whereas Nanegaliños tend to concentrate on community, work and sense of place. Compartmentalisation is more characteristic of the Western-trained modeller's worldview than that of the rural or indigenous community. The connection between people and land transformation is something implicit or hidden in the LUC model but explicit in the local or indigenous framework. While scientists may not agree with local values of production and development, final land use decisions will be made by these land managers whose values and visions of the future will likely be more powerful than distant conservation designs.

Contrasting Scientific LUC with Indigenous Future Visioning: Use of Panoramic Images

After scientific 'derived rules' and scenarios have been ground tested in the cultural context of local people's values, perceptions, and visions of the future, the final step in our methodology was to use this information as a springboard for community-based dialogue about the future. The goal of the 'future visions' methodology is to contrast scientific and indigenous modes of thinking about the landscape and to provide decision makers (at multiple scales and social levels) with insights regarding possible outcomes of different management decisions, whether scientific or locally-derived. By taking this position, we assumed that science is merely one of many valid viewpoints on the future of the landscape and cannot provide prescribed answers for local people. In this view, 'sustainability' is more than a linear biophysical projection based on past trends or 'if-then' simulations. Sustainability is fundamentally a 'societal value about the future' (Röling 1997). Scientists can play a role by understanding how trade-offs or impacts of such values will play out but they cannot determine the values themselves.

A comparative study of maps and landscape images in both Nanegal and Cotacachi led us to the conclusion that panoramic photographs taken from a well-known community viewpoint are more easily recognised and understood by locals than other representations (aerial photos, remote sensing images, cartographic maps, community-drawn maps). Panoramic photo-simulations were thus created for both the Nanegal and Cotacachi landscapes using the following procedure:

1. A 180° panoramic photo, representing the current landscape, was taken from a well-known community viewpoint.
2. Plausible scenarios of the past (e.g., 1950, 1990) were constructed (using the commercial software program Photoshop) based on the land use change maps and information from ethnoecological research and oral history workshops.
3. Alternative scientific scenarios of the future based on the LUC projections were produced to the year 2030 (one as a linear projection and another under conditions of balanced growth).

4. Past, present and alternative images were transferred to large (25 cm × 125 cm) panoramic photographs for easy viewing by local people in a community meeting.

We tested this use of panoramic landscape images using the same procedures in both Nanegal and Cotacachi. Similar to Cree conferences held to confront scientific knowledge, communal gatherings were convened in a central, public location with as many people as possible representing a cross-section of the population. Since we had worked in both areas for five years most participants were acquainted with our project and with us as individuals. Community leaders opened the meetings with a traditional welcome and a discussion on the need to plan for future generations. Our team then explained without technical jargon our idea of future visioning and how we had created our science-based land use projections. We then presented images of the past landscape (e.g., 1950) and asked for comments, especially from the elderly who were the only participants with direct experience who might accurately correct our historic images. We also presented the panoramic photograph of the present landscape, which is an actual photograph from a well-known viewpoint, not a scenario. The images generated a great deal of excitement, debate and corrections compared to stony silence when statistical regression charts were shown. Local people gained confidence because of their superior and intimate understanding of past and present landscapes. Individuals from specific communities were able to make detailed micro-

Figure 12.2 Panorama scenarios for Nanegal for three periods (1950, 2000 and 2030).

landscape comments on the accuracy of our retrospective view. They identified locations of former *haciendas*, named the families who live there today, and discussed details of soils, climate, vegetation and cropping history. The past scenarios elicited intense interest among the elders, who were anxious to correct them. They helped us locate original trails and roads and argued that we had overestimated the past forest cover.

After comparing past and present images, we then presented our landscape photograph simulation of the future based on linear trend projections a generation hence. Figure 12.2 shows the panorama scenarios for Nanegal for three periods (1950, 2000 and 2030). Future scenario (2030) reflects a linear projected growth of population, increasing reliance on income-generating industries such as poultry and floriculture, as well as a disregard for the environment and the projected LUC.

In both Nanegal and Cotacachi, the workshop participants were fascinated by these images of the future. They appreciated that ours was a vision from the 'outside' and it was important that they correct our interpretation as 'insiders' based on their values and needs. In order to further stimulate discussion, we asked the participants to divide into small groups based on gender and age (Jones et al. 1998: 728). This helped us identify intra-group differences in landscape perception and future visioning. Although there was some variation in number of participants between Cotacachi and Nanegal, the visioning groups were: (1) Elders (above age 50); (2) Men and women with productive families (25–49); (3) Young adults, (15–25); and (4) Children below 15. Each group was given a sheet of paper exactly the same size (15 cm × 125 cm) as the panoramic photo containing only perspective contour lines. They were given colour pens and asked to draw the landscape of their community thirty years in the future based on the questions: 'How would you like your community and surroundings to look thirty years from now?' Each group had one hour to complete their vision of the future and prepare a short presentation before the workshop participants.

The future visioning exercise demonstrated that there are significant intra-group perspectives on the future. The very young and the elderly had few specific ideas about how the landscape should look three decades from now. This is most likely because neither thinks they will be there in thirty years, one through death and the other through permanent out-migration to urban centres. In Nanegal, the elderly mainly concentrated on drawing green areas with no infrastructural details except for sewage and water systems. They seemed to be reflecting on the past and their own memories of landscape instead of projecting forward. The young concentrated on their immediate needs: a football stadium, swimming pool, and better roads. The youngsters were more aware of environmental issues and had mastered the right "green" jargon or phrases due to environmental education programmes, but gave little importance to them in their own scenarios.

Nanegal men in their 30s and 40s with families started their future scenarios by drawing infrastructure they already had in the community (church, school, doctor's office, power lines and one community phone). They first sketched the traditional layout and expansion of the town with blocks and squares emanating from the centre of town. They then created a new road system and placed industries according to access to the roads. They then began to design what they would like to have for their children: better roads, high schools and technical schools, sports stadium, sewage system, industries providing employment, monitors of water contamination, trash collection, and hotels. Their main non-infrastructure concerns were institutions that would educate the young and employment opportunities to keep them in the community. The mountain and hillsides around the village were covered with a mosaic of 'ecological farms' where they said families would live, and in between the farms were newly forested areas. They said the main agricultural problem was climate change ('used to rain more'; 'today the sun is more intense').

The Nanegal women's group in their 30s to 40s also concentrated on civil institutions and urban infrastructure during the mapping exercise. Examples of institutions include vocational schools and day-care facilities, which would allow them to become more self-sufficient in Nanegal and less dependent on migration to Quito. When we directed their attention to the territory outside the town centre, the women did not give much importance to forest (and 'biodiversity') or express how the forest might help their opportunities for progress and personal advancement.

The panoramic visioning exercise simply verified what the story completion tests in Nanegal already told us about local values and conception of landscape change. There was not much local interest in preservation of the forest, only in the sense that the forest could help with income by bringing more tourists. Given few employment opportunities in the area, Nanegaliños are making a realistic reading of their future. Those who remain, mainly families of those in their 30s and 40s, are primarily concerned with civil and social infrastructure to improve their lives. It is unlikely that environmental education, biodiversity projects, or biological corridor programmes that mainly employ non-locals will capture the interests of local people. Our modelling focus on deforestation simply did not interest most Nanegaliños, who felt that community building and livelihood support should take precedence over conservation. This local non-interest in forest sustainability had not been envisioned or incorporated into the scientific LUC decision support model.

The results from the future visioning workshops in indigenous communities of Cotacachi contrast with those from Nanegal in significant ways. The Cotacachi LUC model predicted that by 2030 the area of native forest will decline by half compared to 1963, while urban areas will triple in size (Zapata et al. in press). On the other hand, tree plantations (mainly eucalyptus) owned by absentee landlords will increase ten times. Farm parcels less than 3 hectares will double, those of 3–5 hectares will increase more than tenfold while, correspond-

ingly, larger parcels (greater than 5 hectares) will decline substantially. This sig-
nals the continued minifundisation of the landscape due to inheritance patterns
among indigenous people who pass smaller and smaller parcels to their children.
Also, greenhouses for floriculture, although occupying a small land area, will
increase more than 60 percent over the next thirty years. In summary, the
Cotacachi landscape is becoming deforested in the higher elevations, landholdings are continuing to shrink in size, and the urban area – including the floriculture industry – is growing significantly.

When shown the present panorama (2000) and scientific scenarios (1963 and
2030), the Cotacachi elders predictably examined and corrected the 1963 image
(e.g., more snow on the mountain, more forest in some places, etc.). Young people, as in Nanegal, also gave attention to the 2030 image with sports facilities and
public transportation, reflecting not only their interests in travelling to work in
Quito or elsewhere but their youthful social activities. Similarly, Cotacachi family men and women (25–49) emphasised civic and social infrastructure with men
paying more attention to water supply and women to places where social services might be provided, e.g., church, market, and municipality. The need for
socioeconomic advancement was voiced as strongly among the indigenous
Cotacacheños as mestizo Nanegaliños.

In creating their own future scenarios, however, the Cotacacheños projected
a stronger cultural overlay to their landscape and a deeper attachment to ethnic
ties within communities. Reflecting their historical sense of place, they named
their communities, drew Andean symbols of *Pachamama* (Mother Earth) and
Inti (the Sun God), identified community houses, and noted sacred places.
Compared to mestizo Nanegal there was a clear effort to connect their future
with the indigenous pan-Andean cultural revitalisation and political movement.
The future scenario drawings were dotted with statements pointing to issues of
contestation over access to land and the Cotacachi-Cayapas Ecological Reserve.
They specified which modern aspects of the landscape should be removed in the
future ('No high radio/TV antennas', 'No floriculture greenhouses', 'No *haciendas*', 'No urban development'). In viewing the indigenous-created scenarios,
social values of strong cultural identity and attachment to community were more
salient than vegetation cover and land use change. Any effort to addresses sustainable land use without addressing these cultural values will likely be ignored
by Cotacacheños, who proudly claim they want 'development with identity', not
just development (Rhoades 2006).

Another interesting issue arises in the use of scientific linear projections to
model the future landscape in Cotacachi. In this region, as well as throughout
the Andes, the spatiotemporal conception of history (and thus landscapes) is dia-
metrically opposed to Western concepts of linear time as reflected in the projections. For the Quechua, the past is located in front of the viewer, not behind as
in Euro-American time. What you can see out there is your past and that behind
you is the unseeable or the future. To deal with the future means seeing the past
clearly. According to Rappaport, who worked in Colombia,

In Quechua, when appropriate diacritic markers are used, the terms for 'the past' and for 'in front of the observer' are identical – *nawpa*. In other words, in the Andean vision of the past, history is in front of the observer, and moves backwards toward the observer. (Rappaport 1988: 721)

We believe this might explain the greater emphasis in the future vision of Cotacachi on past or present elements in their landscape that they would like to see eliminated in the future. Rappaport (1988: 721) continues:

Their explanation of their particular space-time frame centres on practice as opposed to fact: although events occurred in the past, we live their consequences today and must act upon them now. For this reason, what already occurred is in front of the observer, because that is where it can be corrected. History is, therefore, most relevant to the present and is of the present. History can be corrected, since it is in front of the observer. In this vision of the past, unlike our own, historical correction is not empirical in nature, it does not involve the transformation of historical accounts on the basis of new data. Instead, it is founded on the use value of the past.

Paralleling Sillitoe's (this volume) argument that New Guinea counting systems offer new insights for sustainability, the Quechua view of past and future time, with its infusion of the past into the future, might well be a more sophisticated way to think of sustainability 'a generation into the future'.

Conclusion: a Different Misreading of the Landscape

Our research in Ecuador does not completely bear out the insights found in Fairhead and Leach's (1996) empirically solid and ground-breaking study Misreading the African Landscape. Although the people of Nanegal and Cotacachi, through their subsistence and social activities , have also created new niches of human-induced biodiversity and sustainable production systems (see Rhoades 2001), the scientific modellers have not misread their landscapes in terms of forest loss, which is an inescapable fact. However, our methodology of future visioning does point to a different and equally important misreading. Environmental scientists look at the landscape as an objective, abstract construction based on their formal training and see degradation and need for restoration. Our study reveals that they have forgotten the future of the local people who have other values in mind. Scientists see land use in the abstract and input their own globally informed professional values while people see land use as place and input their local values of survival and community. We would further argue that scientists, much like the people of Northwest Ecuador, would hold similar values of community and livelihood as primary in their own home landscapes.

The methodology and results discussed in this chapter also go to the heart of the debate over conservation versus development in ecoregions characterised by buffer zones and protected areas. The international donor community has called for the systematic integration of local people's needs and values into land management planning for conserving biodiversity. The President of the IUCN (the World Conservation Union) has noted that 'if local people do not support protected areas, then protected areas cannot last' (Ramphal 1993: 56). Research into sustainability makes it clear that defining it only in biological terms without considering the cultural element is inadequate (Obasi 2002). Top-down technocratic approaches have been abandoned in favour of participatory research and action which considers intergenerational justice and equality. However, in most cases, interdisciplinary teams have been hampered by a lack of understanding of how science can be integrated or contrasted with local viewpoints.[3] This is especially true in attempting to envision the future through simulation modelling which relies largely on linear projections of past land use patterns. LUC analysis has largely been carried out independently of local people, the intended beneficiaries of sustainable development.

This paper asks how the LUC modelling exercise compares to the indigenous or local viewpoint about the same landscape and time frame. Local people do not assign the same importance to LUC categories as scientists. However, we have presented a methodology of future visioning to make it possible to subject the 'derived LUC rules' to analysis by local people using ethnoecological methods. In comparing scientific and local perspective on land use change, it is important to understand that we were comparing two different heuristics, or models. These devices aid people in dealing with the complexity and chaos of the world and reducing equivocation so that they can make a decision, a plan for action. While heuristic devices are not intended to capture reality in all its complexity (if possible this would render the models redundant), what is interesting in comparing them is *how each group chooses how to simplify*. This is a critical issue because it reflects on what is considered important information, and therefore worthy of a place in the model, and what is dispensable or superfluous. Another interesting facet is the relationship that is seen among the many aspects of their lives that have changed over the course of three decades and projected to do so over the next three decades.

In the end, we have to realise that 'models are merely representations, useful for guiding further study but they are not susceptible to proof' (Honachefsky 2000: 40). Well-executed scientific models are useful heuristic devices that can somewhat compress time so that we can estimate the future results of a given action. The philosopher, Nancy Cartwright (1983), has claimed models to be 'works of fiction' because a model, like a novel, might resonate with nature but it is not the 'real' thing. Models are useful when they challenge existing ideas and formulations, rather than trying to validate them or verify them (Honachefsky 2000). And they can become more useful when they are systematically compared to the viewpoints of the people whose lives and actions they model.

We think there is an additional value to 'envisioning' with local people beyond the scientific model. We believe it is time to 'go the extra mile' to listen to people's voices and learn about the principles and models they use to navigate and operate in a complex and changing world (Bicker et al.2004). To show them our scientific models and ask them to revise their ways so that pristine environments can be conserved for our aesthetic pleasure, economic needs or our health is paternalistic and counterproductive. They have their own vantage point and their own time frame from which they view and evaluate the changes in land use, soil conditions or hydrological cycles. We need to take their perspectives into serious consideration, or better yet, use them as points of departure, in our effort to arrive at a negotiated future visioning of the landscape.

Notes

1. Our research has shown, in fact, that the frontier around the Cotacachi-Cayapas Ecological Reserve is a stable agricultural area, not a moving colonisation zone. Due to steepness of slope and remoteness of terrain, migrants are not moving towards the reserve but are transforming the areas where they have already settled (Rhoades 2001).
2. The Sustainable Agriculture and Natural Resource Management Project (SANREM) was established by the United States Congress in 1992 as one of the U.S.A.'s responses to meeting Agenda 21. As a Collaborative Research Support Program (CRSP), the programme engaged U.S. and host country university researchers with NGO partners 'to advance the principles, methods, and research, and collaborative breakthroughs for a new paradigm of sustainable development (National Research Council 1991). Three representative landscapes were selected for long-term research: a tropical watershed in Mindanao, Philippines; a semi-arid region of Burkina Faso (later moved to Mali); and two microregions within the Andean buffer zone of the Cotacachi-Cayapas Ecological Reserve of Northwestern Ecuador.
3. Participatory approaches in vogue during the early to mid-1990s had pluses and minuses. Participation by local people satisfied global initiatives such as democratisation and decentralisation. Although more engaging than questionnaire surveys, participatory methods alone (community meetings, conservation circles, debate tables, group treks, etc.) are insufficient to concretely frame ecological or agricultural problems and action at the landscape scale into the future. In community consensus-building exercises, contradictory and conflicting local values and perceptions were not readily resolved due to other influences of local power structure, including the roles of gender and ethnicity. Above all else, however, the community-based interaction process did not generate understandable, empirical or visual information which would allow one to see changes through time and to project plausible scenarios of the future.

References

Abel, N., A. Langston, B. Tatnell, J. Ive and M. Howden nd. 'Sustainable Use of Rangelands in the 21st Century: A Research and Development Project in Western New South Wales'. Manuscript.

Adams, W.Y. 1980. 'Shonto in 1989: a Culture Historian Looks at the Future,' in *Predicting Sociocultural Change, Southern Anthropological Society Proceedings, No. 13*, ed, Susan Abbot and John Val Willigen, pp. 115–26. Athens: University of Georgia Press.

Alarón, R. 2001. 'Biological Monitoring: a Key Tool in Integrated Conservation and Development Projects', in *Integrated Conservation and Development in Tropical America*, (eds), R. Rhoades and J. Stallings, pp. 23–32. SANREM CRSP and CARE-SUBIR.

Ames, S. 1989. *Choices for Oregon's Future: a Handbook on Alternative Scenarios for Oregon Planners*. Oregon Visions Project, American Planning Association (Oregon Chapter), Gresham, Oregon.

Basso, K. 1996. *Wisdom Sits in Places: Landscape and Language among the Western Apache*. Albuquerque: University of New Mexico Press.

Bicker, A., P. Sillitoe and J. Pottier. 2004. *Development and Local Knowledge*. London and New York: Routledge,

Cartwright, Nancy 1983. *How the Laws of Physics Lie*. Oxford: Clarendon Press.

Chocó-Andean Corridor Project. 2000. 'Medium-Sized Project Brief', 9 pp.

Costanza, R. 1997 *Using Envisioning to Design a Sustainable and Desirable World in the Presence of Irreducible Uncertainty*. Paper presented at DeLang-Woodlands Conference on 'Sustainable Development: Managing the Transition'. Houston, Texas, 3–5 March 1997.

Denevan, W. 2001. *Cultivated Landscapes of Native Amazonia and the Andes*. Oxford: Oxford University Press.

DeWalt, B.R. 1994. 'Using Indigenous Knowledge to Improve Agriculture and Natural Resource Management', *Human Organization* 53(2): 123–31.

El-Swaify, S.A. and D.S. Yakowitz, (ed) 1998. *Multiple Objective Decision Making for Land, Water, and Environmental Management*. Boca Raton: Lewis Publishers.

Fairhead, J. and M. Leach 1996. *Misreading the African Landscape: Society and Ecology in a Forest–Savanna Mosaic*. African Studies Series, 90. Cambridge: Cambridge University Press.

Gell, A. 1992. *The Anthropology of Time*. Oxford: Berg.

Gerlagh, R. and E. Papyrakis. 2003. *A Sustainable Future? IHDP Update*. Newsletter of the International Human Dimension Programme on Global Environmental Change. 10–12 January.

Gregory, R. 2000. 'Using Stakeholder Values to Make Smarter Environmental Decisions', *Environment* 42(5): 33–44.

Honachefsky, W. 2000. *Ecologically Based Municipal Land Use Planning*. Boca Raton: Lewis Publishers.

Hyslop, J. 1990. *Inka Settlement Planning*. Austin: University of Texas Press.

van Ittersum, M.K., R. Rabbinge and H.C. van Latesteijn. 1998. 'Exploratory Land Use Studies and Their Role in Strategic Policy Making', *Agricultural Systems* 58(3): 309–30.

Jones, A.C., S.A. El-Swaify, R. Graham and D.P. Stonehouse. 1998. 'A Synthesis of the State-of-the-Art on Multiple Objective Decision Making for Managing Land, Water, and the Environment' in *Multiple Objective Decision Making for Land, Water, and Environmental Management*, (eds), S.A. El-Swaify and D.S. Yakowitz. Boca Raton: Lewis Publisher.

Iroquois Tribe. c.1475. *Gayanahagowa: The Great Binding Law.* Constitution of the Iroquois Nation (available at www.canadahistory.com) 16pp. (quote p. 5).

King, J.L. and K. Kraemer 1992. *Models, Facts, and the Policy Process: the Political Ecology of Estimated Truth.* Working Paper #URB-oo6. Center for Research and Information Systems and Organization (CRITO). Irvine: University of California.

Lambin. E.F., H. Geist and E. Lepers 2003. 'Dynamics of Land-use and Land-cover Change in Tropical Regions', *Annual Review Environmental Resources* 28: 04.1–04.37.

McDonald, K. 1996. 'Crowding the Country: Montana State University. Researchers Study Impact of Urban Refugees'. *The Chronicle of Higher Education,* 24 May.

National Research Council. 1991. *Toward Sustainability: a Plan for Collaborative Research on Agriculture and Natural Resource Management.* Washington, D.C.: National Resource Council.

Nazarea, Virginia D. 1998. *Cultural Memory and Biodiversity.* Tucson: University of Arizona Press.

——— 1999. 'Lenses and Latitudes in Landscape and Lifescapes', in *Ethnoecology: Situated Knowledge/Located Lives,* ed., V. Nazarea. Tuscon: University of Arizona Press.

——— 2005. *Heirloom Seeds and Their Keepers: Counter Memory and Place in the Persistence of Biodiversity.* Tucson: University of Arizona Press.

Nazarea, V., R. Rhoades, E. Bontoya and G. Flora 1998. 'Defining Indicators Which Make Sense to Local People: Intra-Cultural Variation', in *Perceptions of Natural Resources', Human Organization* 57(2): 159–70.

Obasi, Godwin O.P. 2002. 'Embracing Sustainability Science, the Challenge for Africa'. *Environment* 44(4): 8–19.

Ogilvy, J.A. 2002. *Creating Better Futures: Scenario Planning as a Tool for a Better Tomorrow.* Oxford: Oxford University Press.

Peluso, Nancy L. 1995. 'Whose Woods are These? Counter-mapping Forest Territories in Kalimantan, Indonesia', *Antipode* 27(4): 383–406.

Poole, P. 1995 *Indigenous Peoples, Mapping and Biodiversity Conservation.* Biodiversity Support Program, Washington, D.C.

Ramphal, Sir Shridath 1993. 'Comments', in *Parks for Life: Report of the IVth World Congress on National Parks and Protected Areas,* (ed.) J. McNeely pp. 56–8. Gland Switzerland: IUCN.

Rappaport, J. 1988. 'History and Everyday Life in the Columbian Andes', *Man* 23: 718–39.

Rhoades, R.E. 1998. 'The Participatory Multipurpose Watershed Project: Nature's Salvation or Schumacher's Nightmare?', in *Integrated Watershed Management in the Global Ecosystem,* (ed.) Rattan Lal, pp. 327–43. Boca Raton: CRC press.

———— 2001. *Bridging Human and Ecological Landscapes: Participatory Research and Sustainable Development in an Andean Agricultural Frontier.* Dubuque, Iowa: Kendall/Hunt Publishers.

Rhoades, R.E. (ed.) 2006. *Development with Identity: Culture, Community, and Sustainability in the Andes.* Wallingford, Oxon.: CABI Publications.

Rhoades, R., D. Stewart, V. Nazarea and M. Piniero 2000. 'The Sustainable Mountain Future Methodology: an Ongoing Study of Visioning in Nanegal Parish, Ecuador'. SANREM-Andes Working Document 2000–1.

Rhoades, R. and J. Stallings (eds) 2001. *Integrated Conservation and Development in Tropical America.* SANREM CRSP, Athens, Georgia and CARE-SUBIR.

Rhoades, R. and S. Moates 2003. 'Reality 3D: Innovative Representations of an Andean Landscape'. SANREM CRSP Research brief, University of Georgia, Athens, Georgia, (www.iapad.org/publications/ppgis/sanrem.pdf).

RIVM. 1998. *Assessment of the Physical Environment. Structure and Content Put to Trial.* Bilthoven, Netherlands: National Institute of Public Health and the Environment (RIVM).

Röling, N. 1997. 'The Soft Side of Land. Socio-economic Sustainability of Land Use Systems', *ITC Journal* 3/4.

Shabman, L. 1995. *'Shared Vision' Modelling for Environmental Project Planning.* Report prepared for Planning and Management Consultants, Ltd, Carbondale, IL.

Sierra, R. 1996. *La Deforestacion en el Noroccidental del Ecuador, 1983–1993.* Quito, Ecuador: EcoCiencia.

Sillitoe, P., J. Barr and M. Alam 2004. 'Sandy-clay or Clayey Sand? Mapping Indigenous and Scientific Soil Knowledge in the Bangladesh Floodplains', in *Development and Local Knowledge*, (eds), A. Bicker, P. Sillitoe and J. Pottier, pp. 174–201. London and New York: Routledge.

Stewart, D.J. 2001 'Creating Land Use Change Scenarios: Past Patterns and Future Trajectories', in *Bridging Human and Ecological Landscapes: Participatory Research and Sustainable Development in an Andean Agricultural Frontier*, (ed.) Robert E. Rhoades, pp. 170–89. Dubuque, Iowa: Kendall/Hunt Publishers.

T'Selele, F. 1977. 'Statement to the Mackenzie Valley Pipeline Inquiry' in *Dene Nation – the Colony Within*, (ed.) M. Watkins, pp. 12–17. Toronto: University of Toronto Press.

Talawar, S. and R.E. Rhoades 1998. 'Scientific and Local Classification and Management of Soils', *Agriculture and Human Values* 15: 3–14.

Tebtebba, Fourdafon 2000. 'We Keep the Past not Behind us But in Front of Us!' *Marila Declaration of the International Conference on Conflict Resolution, Peace Building, Sustainable Development, and Indigenous People.* Organised and convened by Tebtebba foundation (Indigenous People's International

Centre for Policy Research and Education) in Metro Marila, Philippines. 6–8 December.

Textor, R. 1999. 'Why Anticipatory Anthropology?' *General Anthropology* 6(1): 1–4.

Twery, M.J., S. Stout and D. Loftis 1998. 'Using Desired Future Conditions to Integrate Multiple Resource Prescriptions: the Northeast Decision Model', in *Multiple Objective Decision Making for Land, Water, and Environmental Management*, (eds), S.A. El-Swaify and D.S. Yakowitz, pp. 197–203. Boca Raton: Lewis Publishers.

United Nations Conference on Environment and Development (UNCED). 1992 Agenda 21: Programme of Action for Sustainable Development. Rio Declaration on Environment and Development. United Nations Publication.

U.S. Department of Agriculture. 1995. *An Introduction to Image Processing for Visual Simulation*. Natural Resources Conservation Service.

Wallace, Anthony F.C. 1966. Religion: an Anthropological View. New York: Random House.

Wollenberg, E., D. Edmunds and L. Beck. 2001. 'Anticipating Change: Scenarios as a Tool for Increasing Adaptivity in Multistakeholder Settings,' in *Biological Diversity: Balancing Interests Through Adaptive Collaborative Management*, (eds), L. Buck, C. Geisler, J. Schelhas and E. Wollenberg. Boca Raton: CRC Press.

World Commission on Environment and Development. 1987. *Our Common Future*. Oxford and New York: Oxford University Press.

World Future Society. 2002. *The Futurist: a Magazine of Forecasts, Trends, and Ideas about the Future*, Sept–Oct., 36(5).

Zapata, X., R. Rhoades, M. Segovia and F. Zehetner 2006. 'Four Decades of Land Use Change in the Cotacachi Andes: 1963–2000', in *Development with Identity*, (eds). R. Rhoades, pp. 46–63. Wallingford Oxon: CABI Publications.

Counting on Local Knowledge

Paul Sillitoe

In New Guinea, people do not count far. Some languages only have two words for numerals – one and two – and some have no words for numbers at all. It will probably strike persons reading a collection of papers originating at a British Association for the Advancement of Science annual festival to celebrate and advertise the achievements of science, with its associated sophisticated mathematical logic and computational power, as ridiculous to suggest that such numerical schemes can teach us anything. But that is exactly the sort of thing that those of us advocating attention to local knowledge are arguing, particularly in development contexts. The contributions to this book intend to convince you that such advocacy is not as daft as it may at first appear. In arguing this position, this chapter takes a cursory look at the approach science adopts to measurement, particularly the numerical scheme it uses to record and manipulate findings.

Science relies heavily on quantification and mathematics comprises an important part of its language. 'Nature speaks in equations. The rules of mathematics … govern the way the universe works' (Seife 2000: 117). Lord Kelvin, immortalised in the absolute temperature scale, put it clearly in his observation that, 'When you can measure what you are speaking about, and express it in numbers, you know something about it; but when you cannot measure it, when you cannot express it in numbers, your knowledge is of a meagre and unsatisfactory kind: it may be the beginning of knowledge, but you are scarcely, in your thoughts, advanced to the stage of science' (quoted in Seife 2000: 158). I do not think that such understanding is meagre at all. It has something to tell us. The numerical system science uses has unresolved mysteries at its heart, highlighted by the polar notions of zero and infinity, which possibly compromise the idea of sustainable science and development. When science is harnessed to the demands of the market economy, as has increasingly happened, the idea of sustainability becomes even more remote given economics' apparent aim of endless growth. It is possible that those who count less have some important conceptual lessons to teach us about sustainability.

Can We Count on Numbers?

The manipulation of numbers has proved so powerful that it goes largely without question. But what is the character of the reflection of reality that the way we count gives us, which underpins those notions of measurement that are central to scientific experimentation to 'prove' the 'validity' of hypotheses? This is an old question that goes back at least to the ancient Greeks, who questioned the partial, even distorted view of the world that our counting results in. Zeno captured it in his famous paradox questioning the commonsense idea that we live in continuously structured, infinitely divisible space (Rucker 1995: 78–82; Seife 2000: 40–48; Salmon 2001), what is called the continuum problem. In order to make one step he reasoned, we have to complete an infinite number of actions. Any journey starts with one pace of a metre or so, and that single step begins by covering one centimetre, and that starts with one millimetre, and that by covering one micron, and so on ad infinitum. The idea that every action comprises sub-actions that consist of sub-sub-actions and so on, such that before we can move we have to complete an infinite number of ever smaller consecutive actions, flies in the face of our everyday experience – after all, you made it out of bed to wherever you are now reading this book. The alternative view is that space exists of discrete units and can only be represented using indivisible 'whole numbers', but this view is as perplexing as the infinite units one that represents space using endlessly diceable 'real numbers' (Horgan 1996: 231). In the discrete unit view there are only a finite number of points between two places but in this event how can we get from one to another because there is nothing between them? There is nowhere to be while crossing the gaps between the points, which is to be in nothingness passing through a void.

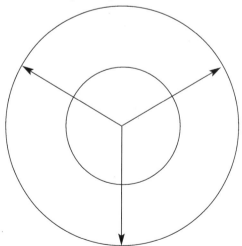

Figure 13.1 The concentric circles of Scotus.

Zeno's paradox suggests that we can infinitely divide any line, Greek geometry assuming that lines are infinitely extensible and divisible. John Duns Scotus, a medieval Scottish philosopher, argued that if a line comprises an infinite number of infinitesimal points, then any concentric circles should comprise equal numbers of points, as a diagram pairing off points intimates (Figure 13.1). The implication is that the two circumferences should be the same size, which they are not. Galileo subsequently maintained that the larger circle comprises infinitely many infinite gaps not found in the smaller one, something that our finite minds cannot grasp, leading us to contradictions. Nicholas of Cusa further showed that finite mathematical distinctions collapse at the level of the infinite; for example, the curvature of a circle's circumference diminishes with size such that we might surmise that an infinite circle would comprise an infinite straight line. They all argued that our inability to conceive mathematical infinity reflects our finite minds. Such comparisons of size make no sense in infinite terms (Moore 1990: 50–55).

These conundrums relate to the mystifying polar concepts of infinity and zero that have puzzled thinkers for millennia (Seife 2000; Barrow 2000). They are central to our conception of numbers, and 'it is hard to see how mathematics could exist without the notion of infinity, for the first thing a child learns about mathematics – how to count – is based on the tacit assumption that every integer has a successor' (Maor 1991: ix). Mathematics has been described as the science of the infinite, in that it presupposes an infinite framework (Benardete 1964: 263; Moore 1990: 147). Yet, paradoxically, 'When you have an infinity in an expression, or when you divide by zero, all the mathematical operations – even those as simple as addition, subtraction, multiplication, and division – go out the window. Nothing makes sense any longer' (Seife 2000: 129). We all learn at school that zero does strange things in mathematical operations, for example that dividing by zero destroys the fabric of mathematics by allowing us to prove anything (Seife 2000: 23), including the existence of God as in Huxley's commentary:

> You know the formula: m over nought equals infinity, m being any positive number? Well, why not reduce the equation to a simpler form by multiplying both sides by nought? In which case you have m equals infinity times nought. That is to say that a positive number is the product of zero and infinity. Doesn't that demonstrate the creation of the universe by an infinite power out of nothing? (Huxley 1928: 135)[1]

Scientists have largely chosen to put these baffling concepts of forever and nothing to one side as they have pushed on, using the mathematics of which they are inevitable aspects to 'verify' their formulations about the world. But as Hilbert (1964: 136) noted many years ago, 'the definitive clarification of the nature of the infinite, instead of pertaining just to the sphere of specialized scientific interests, is needed for the dignity of the human intellect itself'. The advent of calculus and later quantum theory were thought to have got round these problems. The idea of the limit and approaching zero in calculus may allow surprisingly accurate approximations and predictions, as scientific technology evidences, but it does not resolve the zero/infinity paradox.[2] The theory of the atom where observable quantities have discrete values and motion is continuous has given rise to the irresolvable idea of particle-wave duality where the continuous and discrete exist simultaneously, again a fudge that does not sort out the zero/infinity mystery. It remains like a flaw in the foundations of contemporary science, suggesting its limited scope and partial view of reality. Infinity causes endless problems in astrophysics. Until recently the 'big bang' theory ignored it, maintaining that what happened before this event is unknowable, but some are rethinking cosmological theories, postulating the existence of infinite parallel universes. Within a universe observable quantities are discrete, while the multiverse comprises a continuum and transitions do not occur entirely in one universe. Astronomers are also questioning the existence of the vacuum as something where nothing is present at all, that is a state of zero energy, currently defining it as 'only a state of minimum energy' (Barrow 2000: 242). At the other end of the scale, zero is a problem in sub-atomic physics regarding the nature of exotic particles such as hadrons, quarks, mesons and so on, which exist as mathematical constructs.[3] What might comprise these, and might we not expect to divide these sub-sub-atomic entities again – as 'an endless succession of smaller and smaller particles, nestled inside each other like Russian dolls (Horgan 1996: 61) – for it is difficult to conceive of something that cannot be divided again, or if divided becomes two lots of nothing? In this view, science appears to be the study of the eternal division of the infinite.

The set theory of Cantor attempts to explore the paradoxes of infinity, but its formal procedures, while logically consistent internally, scarcely settle them. It demonstrates, for example, that a subset of an infinite set will have the same number of elements as the parent set, all corresponding to a transfinite number (*aleph*, denoted χ) 'beyond the infinite' (Love 1994: 659; Maor 1991: 54–60; Moore 1990: 118–28) – i.e. the elements comprising the infinite set of natural numbers equal the elements of the subset of even numbers, odd numbers, composite numbers, or prime numbers etc. (all= χ^0). But this challenges our finite intuition: how can a subset have the same number of elements as the entire set? We can see this by asking what we should expect to be the larger, the infinite series of all natural numbers (1, 2, 3, 4, 5 etc.) or of even numbers only (2, 4, 6, 8, 10 etc.)? Furthermore, although any infinite subset of a set series will have the same infinite *aleph* transfinite number, not all sets have the same number; the 'continuum' transfinite number, for example, applies to points on lines, two-dimensional plane figures such as squares, circles and polygons, and three-dimensional space, such that the number of points in the intervals [0,1] and [0,2] correspond to the same transfinite number (all= χ_1) – indeed there are an infinite number of different infinities, not one infinity (Moore 1990: 147–58; Love 1994: 666). Insistence on the infinite results in endless quests – even when we seem to have reached thresholds – for example, although no one knows how to discover the next prime number, faith in Euclid's demonstration two millennia ago that there are an infinite number of them,[4] means that the search is on to find the next prime and the next one etc.[5] (Sabbagh 2002; du Sautoy 2003). It is perhaps salutary to recall that poor Cantor finished his days in an institution for the insane and that some mathematicians continue to think his ideas misguided.

Counting in New Guinea

There are other ways to count that do not encounter the conundrums of zero and infinity, which do not take 'the infinitude of the counting numbers as an axiom' (Maor 1991: 40). The Wola of New Guinea have very clear ideas of quantification and may spend long periods engaged in calculations, particularly regarding the transaction of wealth in the sociopolitical exchanges that feature centrally in their culture.[6] It is the exchange of wealth, a complex institution, and not corporate nor faction group obligations, that informs their acephalous political environment, while simultaneously allowing individuals comparative *political* sovereignty (Sillitoe 1979). People transact wealth including pigs, sea-shells, cosmetic oil, and today cash, with one another in the interminable series of sociopolitical exchanges that mark all important social events like marriages and deaths, carefully enumerating the wealth presented in public displays. Interestingly, Biersack (1982) interprets Paiela counting in such exchange contexts as facilitating communication between kin on the state of their relations, not as counting per se. The Wola have never intimated such to me, nor their behaviour suggest it. They value exchange highly, according high status to those who excel at it. It is even possible sometimes for these more able men, by virtue of their admired success, to exert a marginal degree of influence over decisions reached by those united for some purpose. The transactions remain today a significant force for order in their fiercely egalitarian, sometimes violent society with non-existent central government authority and lawless 'rascal'[7] activity prevalent throughout the region. Another more recent context in which the Wola show great interest in counting and facility at arithmetic is when gambling with playing cards, a popular activity that can also explode into violent disputes.

The Wola number system counts up to forty-four, or forty-eight when counting in pairs.[8] It is a base four or quaternary counting system,[9] the first eight numbers standing alone, after which numbers are grouped in fours (Table 13.1). The counting scheme is the reverse of the English decimal one, where we add numbers after any decimal class marker and count on, for example, from twenty to twenty-one, twenty-two etc. The Wola do the reverse and group together the previous three numbers of a foursome, so twenty is suwpuw but *suwpuwnmond* (literally *suwpuw's* one) is seventeen, *suwpuwnkab* (literally *suwpuw's* two) is eighteen and *suwpuwnteb* (literally *suwpuw's* three) is nineteen.[10] If someone reaches forty-four and wishes to count beyond this number, he refers to the

Table 13.1 The Wola counting system.

Wola	English	Wola	English
mond	one	*dowapuwnteb*	twenty-three
kab	two	*dowapuw*	twenty-four
teb	three	*seraypuwnmond*	twenty-five
mak	four	*seraypuwnkab*	twenty-six
suw	five	*seraypuwnteb*	twenty-seven
dowa	six	*serayapuw*	twenty-eight
hogo	seven	*bekenpuwnmond*	twenty-nine
dukab	eight	*bekenpuwnkab*	thirty
tobonmond	nine	*bekenpuwnteb*	thirty-one
tobonkab	ten	*bekenpuw*	thirty-two
tobonteb	eleven	*halanpuwnmond*	thirty-three
duteb	twelve	*halanpuwnkab*	thirty-four
mogapuwnmond	thirteen	*halanpuwnteb*	thirty-five
mogapuwnkab	fourteen	*halanapuw*	thirty-six
mogapuwnteb	fifteen	*kompuwnmond*	thirty-seven
mogapuw	sixteen	*kompuwnkab*	thirty-eight
suwpuwnmond	seventeen	*kompuwnteb*	thirty-nine
suwpuwnkab	eighteen	*kompuw*	forty
suwpuwnteb	nineteen	*iygaypuwnmond*	forty-one
suwpuw	twenty	*iygaypuwnkab*	forty-two
dowapuwnmond	twenty-one	*iygaypuwnteb*	forty-three
dowapuwnkab	twenty-two	*iygaypuw*	forty-four

things enumerated as one lot of forty-four, and starts counting again. Instead of operating with an infinitely extendable and divisible number line, he counts in rounds, starting a new circuit at one when he reaches forty-four. In theory a person could count up to forty-four lots of forty-four (i.e. 1,936), and I suppose he could even go beyond by calling this sum one lot and starting again (i.e. go beyond 1,936), but no one ever has reason to count to such large numbers, which would be to take us in the direction of infinity, a concept that I have heard no one allude to and for which, so far as I am aware, there is no word. The nearest are words for 'very large' (*onda ora*) or 'very many' (*onduwp ora*), which people may employ in the sense 'there are too many to count', such as ants on a nest, but not to mean infinite in number. Other counting systems in Melanesia likewise reach an indeterminate ceiling. As Wolfers (1972: 219–20) notes: 'There was not so much a limit to counting as a limit to the goods and quantities that needed to be counted or that particular groups of people wanted to measure.' The Iqwaye speakers, for instance, resort to 'many' beyond a certain point when the number of things become so numerous as to be too daunting or worth counting, in the sense that enumerating them would not serve any useful purpose, for the 'the Iqwaye counting system does not aim to generate the numeri-

cal series as an indefinite, linear and ultimately open succession of units'
(Mimika 1988: 73).[11] The Maenge of New Britain resort beyond four hundred
to metaphors of numberless, such as like 'the leaves of a tree' or 'the sands on the
beach' (Panoff 1970: 362). Similarly, the Buang, for whom four hundred indicates 'a large indeterminate number', may refer to fallen tree leaves when referring to large numbers (Hooley 1978: 155).

While the counting systems may theoretically go on to very high numbers,
albeit with increasing cumbersomeness, why should people want to do so? Here
it is helpful to make a distinction between number applied to concrete things
(such as pigs or money) and number manipulated as abstract symbols (as in
mathematical logic; see Thune 1978: 74, Biersack 1982: 813; Mimika 1988). An
Inuit myth catches the futility of going to very large numbers in the first context, telling of two hunters entering into a contest to settle an argument over
whether a caribou or a wolf has the most hairs, and becoming so engrossed in
pulling out hairs one at a time and counting them that they starved to death –
the moral being that this 'is what happens when one starts to do useless and idle
things that can never lead to anything' (Seidenberg 1962: 33). Humans have
devised many ways to depict large numbers without invoking the impossible idea
of dealing with the infinite. One is the tower of Hanoi, which comprised three
poles, on one of which were stacked sixty-four discs of decreasing diameter. The
monks said that if they managed to move all the discs to another pole without
ever placing a disc of larger diameter on a smaller one, the world would end. This
would take 2^{64-1} moves, which at the rate of one move per second, would take
just over 584,942 million years.[12]

In Melanesia we find a wide range of counting systems, in keeping with the
region's fabled cultural diversity (Laycock 1975; Lancy 1983: 102–10). Several
are similar to the Wola scheme. Among Highlanders, for example, we find the
neighbouring Huli to the west have words for numbers up to fifteen (Cheetham
1978), the Kewa up to twenty (Franklin and Franklin 1962), twenty-four among
the Kakoli (Bowers and Lepi 1975), twenty-eight among the Paiela (Biersack
1982), and sixty among the Kapauku (Price and Pospisil 1966), which these people can manipulate like the Wola to count to higher numbers if necessary. Others
are minimalists and have only two words for numbers – one and two- which they
may combine in various ways to count beyond two, such as the Kiwai speakers
(Smith 1978: 55–56), the Iqwaye (Mimika 1988) and the Umeda (Gell 1975:
162). The Iqwaye, for example, count to four using combinations of their one
and two terms (one, two, one-two, two-two) and denote five as one hand,
beyond this they count in groups of fives, using hands and feet, one person signifying twenty, and above this they count in persons and digits (e.g. two persons
and three digits equals forty-three – Mimika 1988).[13] Several such vigesimal systems have been reported in the region (Wolfers 1972: 217; Laycock 1975: 222,
224), including the Maenge (Panoff 1970), Buang (Hooley 1978) and
Normanby Islanders (Thune 1978).[14] Finally, there is the ultimate avoidance of

the zero to infinite real number conundrum: some languages have no words for numbers at all. In these systems people count by referring to parts of the body, such as the Foi, living to the south of the Wola around Lake Kutubu (Figure 13.2), who start counting on one hand and work up the arm across the face and down the opposite arm to the other hand. Each of their number terms is the same word for the corresponding part of the body such that *bonagi* 'wrist' refers to 'seven' and 'thirty-one' and *kia* 'ears' refers to 'fifteen' and 'twenty-three'; the Foi count up to thirty-seven using this system (Williams 1940: 33–34). The Oksapmin have a similar body part counting system that goes up to twenty-seven (Saxe 1981: 307), the Imbonggu one that extends to sixty-eight (Pumuge 1975: 23) and the neighbouring Kewa to forty-eight (Franklin and Franklin 1962: 188), the Daribi scheme extends to thirty (Wagner 1967: 245), and the Paiela have a partial version covering numbers six through to twenty-two (Biersack 1982: 813).[15]

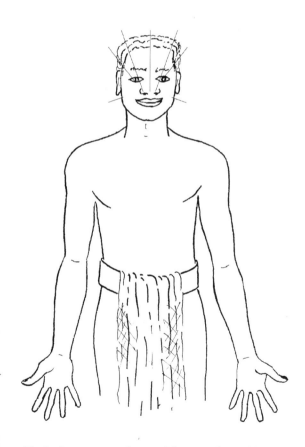

Figure 13.2 The body counting scheme of the Foi (after Williams 1940)

When counting large numbers, the Wola may count in pairs, using a special number marker *hiyp* (e.g. *hiyp teb*, which is '*hiyp* three', is three pairs or six).[16] When they reach twenty-four pairs, they denote this one lot of forty-eight with another special number marker *hort* and start over again counting from one pair until they again reach twenty-four pairs (e.g. *hort kab*, which is '*hort* two', is two lots of twenty-four or forty-eight). Someone could theoretically count up to 2,304 using this system, or referring to a *hort* of a *hort* even higher. They count cowrie shells in this manner in pairs (*hiyp hogo* = fourteen shells, *hiyp suwpuw* = forty shells etc.), and when they reach twenty-four pairs they call this hort mond and start counting from one pair again – a necklace of 118 shells will be *hort kab hiyp tobonteb* (that is, two lots of twenty four pairs and eleven pairs). Such counting in pairs may focus on even numbers but the Wola do not, to my knowledge, distinguish verbally between odd and even numbers (Panoff 1970: 358 makes the same observation for the Maenge). If a necklace comprises an odd number of shells, the counter refers to the last unpaired shell as *ponay* 'free', that is a shell not counted but given for nothing.

When counting to large numbers, people sometimes resort to using small lengths of stick called *iysh hul* 'wood bones' as counters, often arranging them on the ground in front of them. Others regularly use their digits. They may count off their fingers on either hand, pointing to them, bending them over or straightening them up, and finishing with the two thumbs to make ten, sometimes tapping their fists or clapping their palms together to signify ten (similar to the Melpa; Strathern 1977: 16). If sitting down, they may extend the count to the toes, up to twenty. Some persons may use other parts of the body too; after counting the fingers they may count up the arm and across the face or chest and down the other arm. There is no conventional order to this torso counting, as reported for some regions (Williams 1940: 33–34; Franklin and Franklin 1962; Pumuge 1975, 1978; Saxe 1981). The parts of the torso and head counted vary between individuals, as they do among the neighbouring Huli (Cheetham 1978: 24) and also the Paiela (Biersack 1982: 815). The words the Wola use for numbers are not the same as those they use to refer to certain parts of their anatomy, as noted above for some people such as the neighbouring Foi; they are numerical abstractions.[17] Some persons may use this way of counting as a mnemonic device for accounting purposes, such as reckoning wealth owed to them in a forthcoming exchange transaction, remembering the number as reaching the forearm or nose. It is also common to hear people referring to a sum as 'one man' (*ol iysh mond*), which may equate with forty-four, that is counting the digits and various parts of the body up to this number to give 'one man'. Today it is more usual for this phrase to refer to twenty, that is the sum of an individual's digits, reflecting the switch that is occurring to the European base ten counting system with the arrival of cash (ten lots of ten toea coins [previously Australian cents] comprising one kina [previously dollar], and kina notes coming in two, five, ten and fifty note denominations).[18] Another counting device used to record a number for reference purposes is a length of string with knots tied in it, called *pong liy* ('knot hit'). The knots may represent moons, for example, and as each passes the owner undoes one until none are left when he arrives at the agreed date, for an exchange payment or whatever. Or knots may represent reparation payments owed for relatives killed fighting in hostilities sparked off by a dispute in which other kin were embroiled whom the deceased went to support, the knots again untied one at a time as the payments are received (the Iqwaye had a similar knotted mnemonic device, among several other elaborate procedures – Mimika 1988: 14–17).[19]

At the other end of the scale, the Wola do not proceed far in dividing up one. If they split something such as a tree trunk into two halves they may refer to either half as *genk mond* ('small one'). If they divide these halves in two again to give quarters they are *giy mond*. And if they divide these quarters further, the pieces are *tongok*, no matter how many there are. If the pieces split off are small, they are called *dekel* (splinters). Again, with fractions, we have a four base system with terms down to one-quarter (*giy mond*), after which we have the possibility of an indeterminate number of pieces.[20] No one has ever suggested to me the idea of dividing something up endlessly into ever smaller pieces, nor referred to zero in a numerical context. They do, however, express ideas of nothingness in the sense that something may be used up – *orasha* – such as no water left in a gourd drinking vessel (*iyb orasha* 'water finished'), or that something is not present – *na wiy/na hae* – such as there are no pigs in my homestead (*showmay na hae* 'pigs not stand'). And opposed to the idea of very large or very many they refer to very small (*genk sha*).

The Wola have no graphic symbols for numbers nor units of measurement for length, mass and so on,[21] which are central to scientific practice, and from which European mathematics evolved measuring land holdings on the Nile floodplain and store house contents in Mesopotamia, passing from Egyptian hieroglyphs and Babylonian cuneiform to Brahmi-informed Arab notation (Swetz 1994; Barrow 2000: 19–45). While the Wola can talk about size and so on, they do so using relative concepts. Something can be 'long' (*sol*) or 'short' (*hiy*), or qualified versions of these terms, such as 'very long' (*sol ora*), 'longish' (*solsha*) etc. Similarly, regarding weight, something may be 'heavy' (*jendbiy*) or 'light' (*tiyshabiy*), and these words can likewise be qualified, but not quantified. People may also use comparative markers, such as 'hand sized', 'forearm long', 'tall as a man', 'heavy as an axe', and so on.

In contrast to units of measurement, the language does feature markers for different things enumerated, so-called numeral classifiers (Ascher 1991: 10). These extend on the *hiyp* and *hort* pair markers, mentioned above. When counting round things people refer to '*hond* so many' (such as twenty coins, *hond suwpuw*), live creatures are iysh (such as nine pigs, *iysh tobonmond*), and periods of time are uwk (such as six moons, *uwk dowa*). There are seven such number markers for different classes of things (Table 13.2). There is no readily discernible

Table 13.2 Number markers for different classes of things.

Number marker	Used for/precedes	Examples
hiyp	Things counted in pairs/divisible in two	Pandan nuts [*hiyp suw* = five nuts divisible into ten halves]
hond	Round things	Pumpkins, earth mounds, coins
hort	Pairs of things (up to twenty-four)	Cowrie shells, dogwhelk shells
iysh	Live creatures	Human beings, pigs, marsupials
pa	Areas, flat things	Gardens, bank notes
tok	Irregular-shaped things	Pearl shells, string bags, axes, items worn
uwk	Time periods, structures, some plants	Days, moons, houses, bridges, trees, sweet potato

principle underlying the numeral markers, although some of them have a geometrical reference. The neighbouring Huli employ a similar system of markers (Cheetham 1978: 20–22), and the more distant Ponam on the island of Manus (Lancy 1983: 108).[22] Scientists will be unable to conceive how anyone could undertake scientific enquiry with such mathematical ideas, or the lack of them, with no graphical figure notation (signs for 1, 2 etc.) and no units of measurement. And this is the point: the cultural positioning of knowledge determines our understanding of issues.

The Wola system is, as indicated, just one of the several hundred that exist in New Guinea, and there are many others around the world that are quite different to the Euro-Asian one. They give us a different perspective on the problems that we encounter with science, where something seems to be fundamentally wrong, leading the world in unsustainable directions. We are in good company in drawing attention to counting systems that eschew the infinite. The Greeks were so hostile to the idea of the infinite that legend says the Pythagorean brotherhood drowned Hippasus, one of their number, for divulging their problem with irrational ratios (Seife 2000: 37–38). And Aristotle tried to resolve the problem with his ingenious distinction between the actual, as opposed to the potential mathematical infinite. Several eminent philosophers, such as Brouwer, Dummett and Wittgenstein, have subsequently argued that we should be more critical of our mathematics' elementary assumptions. According to Moore (1990: 222), 'our concept of the mathematically infinite is just like our concept of the metaphysically infinite: it has no possible direct application to reality'.

Circular Perspectives

It will seem absurd to scientists to suggest that a tradition in which people only have words for eleven groups of four numbers and until recently men used stone tools can teach us anything – and indeed it appears unlikely that New Guinea Highlanders, left to their own devices would, in the foreseeable future, have learnt to fly, make atomic weapons or transplant organs. But that does not imply that we have nothing to learn from them, or worse, that we should look down on them. Darwinian ideas of evolution have long encouraged this view of other cultures but anthropologists argue against them because their research has taught them that it is dubious to talk about evolution in sociocultural contexts. The world comprises many parallel cultural universes. None knows the truth. All have a view (see Heckler, this volume, for an interesting example of how the imposition of scientific quantification on Amazonia Indians distorted their views). When judged according to certain criteria some appear more 'advanced' than others. According to technological achievements and ability to manipulate nature, Euro-American scientific culture is undoubtedly premier league. But if we consider sustainable lifeways, or spiritual fulfilment or family values, this culture is second division or worse. We cannot meaningfully rank societies evolutionally from primitive to advanced. We cannot see into the future and predict evolutionary outcomes. It is conceivable that Western society will not prove the best adapted or most advanced social formation but that science and the technology it spawns will fatally compromise those who depend on it with pollution and despoliation, social disarray and use of horrendous weapons of destruction. It is an unsustainable trajectory.

The theme of the British Association Festival from which this book of essays originates was sustainable science, and doubtless science can contribute to tackling problems evident regarding the unsustainable nature of much industrial technology. But, we argue, it will help to draw on other knowledge traditions to guide the science giant's footsteps. New Guinea Highlanders were among the first people in the world to cultivate crops. They have evolved environmentally well-adapted agricultural systems, and as farmers have sustainably exploited their montane environment for some 9,000 years. The land and its health are culturally salient to them. Sustainability is deeply ingrained here. The limited counting system intimates a finite perspective of the world, a view reinforced by the limited regions that language groups occupy; the Wola region comprises five valleys, which before contact literally made up the known world. A finite view encourages sustainable attitudes; one has to care for what there is and not squander resources with reckless abandon, exploiting them as if they are endless. This complies with Hilbert's (1964: 151) conclusion that the infinite 'neither exists in nature nor provides a legitimate basis for rational thought' (although he drew on evidence subsequently questioned from quantum mechanics and set theory).

The Highlander way of counting in circles, with no zero or infinity, features closure, albeit indeterminate. In real life the closed system comprises a dense endless network (although no one has ever described it as such to me, I surmise this from my experience of their behaviour). The scientific view is more akin to a line that has no beginning or end, with zero and infinity at the poles, both of which are beyond reach (chaos and complexity theory are now challenging this approach to a remarkably complicated world, which has sought to reduce nature to a series of experimentally tested hypotheses expressed in simple equations). If we follow around a network we trace out a circuit, albeit interconnecting with many other circuits. Shifting cultivation, for example, presupposes a circular movement at the core of this land use system, a complex pattern with farmers coming back to plots after some years of fallow abandonment. In Highlanders' exchanges, wealth flows interminably in space and time around dense networks in ceaseless transactions. A belief in ancestor spirits, called *towmow* by the Wola, who eventually meld into a greater spirit force that animates supernatural and natural life, is another circular conception; one that others, such Aborigines, have elaborated on greatly with their Dream-time ideas. These are inherently sustainable views. In a perceptive challenge to the dominance of lineal conceptions, Lee (1950), for example, argues that for the Trobriand Islanders 'value lies in sameness, in repeated pattern, in the incorporation of all time within the same point' such that their accounts of time seek 'the undisturbed incorporation of events within their original, nonlineal order' (1950: 95), in an attempt to keep the world stable (see also the contribution by Rhoades and Nazarea to this volume on Andean Indian non-linear conceptions).[23]

Any closed structure overcomes the problem of the endless line. The simplest no-end structure is a circle: 'Since times immemorial the circle has been the symbol of regular occurrence, of periodicity' (Maor 1991: 68). It features widely in Asian philosophies in paradoxical plays with opposites, having no beginning and

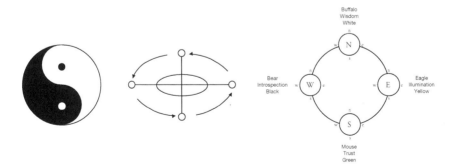

Figure 13.3 (a) The T'ai-chi T'u, (b) Kongo cosmogram and (c) medicine wheel (after Croal and Darou 2002)

end problems. Some cultures have sought to elaborate on these ideas graphically (Figure 13.3), such as the well-known *T'ai-chi T'u* of the Chinese, uniting the polar opposites of *yin* and *yang* – the one representing the feminine, earthly and intuitive and the other the male, celestial and rational. The Kongo use a less known cosmogram that shows human souls in continuous circular motion, life and death mirroring the sun's rising and setting cycle – the living above the horizontal line meet the ancestors below it, while the vertical line links 'above' and 'below' worlds of the living and dead, the Deity and humans (Janzen and MacGaffey 1974: 34; Thompson 1983: 109). Another more familiar depiction is the North American Indian medicine wheel (Croal and Darou 2002: 84–86). The renowned Black Elk pointed out that

> everything an Indian does is in a circle, and that is because the Power of the World always works in circles, and everything tries to be round. … The sky is round, and I have heard that the earth is round like a ball, and so are all the stars. The wind, in its greatest power, whirls. Birds make their nests in circles, for theirs is the same religion as ours. The sun comes forth and goes down again in a circle. The moon does the same, and both are round. Even the seasons form a great circle in their changing, and always come back again to where they were. The life of a man is a circle from childhood to childhood, and so it is in everything where power moves. (Neihardt 1961: 199)

Some Western philosophers have turned to the circle too. It was the geometrical image that Hegel used for infinity, not an endless straight line (Moore 1990: 100). And Blaise Pascal captured nature's awesomeness when he described it as 'an infinite sphere whose centre is everywhere and circumference nowhere' (Moore 1990: 76). But when Western science considers such closed structures, it returns us, with its passion for measurement, to never-ending calculations again, this time via irrational numbers (Rucker 1995: 108–17). We have incommensurable ratios, epitomised in the ratio of the diagonal of a square to its sides [$\sqrt{2} = 1.41\infty$], the 'golden rectangle' ratio phi ($\phi = 1.61\infty$) with its associated 'nautilus shell' logarithmic spiral diminishing infinitely to a zero point, and the ratio pi of the circumference of a circle to its diameter ($\pi = 3.14\infty$).[24]

Infinite Economics

The way we count informs other aspects of our worldview, beyond the scientific. The belief that we can measure everything is an integral aspect, for instance, of our market mentality. Economics relies heavily on quantification, and consequently prides itself on being the most scientific of the social sciences. It also promotes the unsustainable philosophy of endless economic growth, measured as gross domestic product or whatever. It appears that we can never be rich enough; our demand for money and what it purchases is infinite. The close association of scientific research with capitalist economics – currently promoted by politicians and others seeking to link science funding to big-business partnerships – further biases Euro-American understanding and exploitation of the world in unsupportable directions. The market model not only subscribes to the specious idea of infinite growth but it also promotes other falsehoods. We are told that market competition efficiently sorts out resource allocation and consumption. This assumes perfect market conditions, when few – if any – economists think that such conditions exist. They are an ideal of economic philosophers, assuming equal knowledge and information, access to resources and capital. The markets of the West are grossly imperfect with monopolies, cartels and so on, and to assume otherwise is duplicitous or ignorant, which may suit spin-doctoring politicians but 'truth-seeking' scientists ought not to tolerate it.

The harnessing of these market assumptions to contemporary science magnifies the 'errors' of the dominant Western view of the natural and social worlds and their relations. The combination of natural science with capitalist economics is an unholy alliance resulting in today's market science, encouraging an unsustainable mentality. We have wanton pollution because the market fails to make polluters responsible, instead encouraging them to damage our commons because it pays to do so and they can get away with it. Other cultures with more finite views, reflected in circular counting and being, surely have something to teach us. The thought of handing on a degraded natural inheritance is a sin to many tribal peoples, as conveyed in the following observation of a Californian Indian:

> The White people never cared for land or deer or bear. When we Indians kill meat, we eat it all up. When we dig roots we make little holes. When we built houses, we make little holes. When we burn grass for grasshoppers, we don't ruin things. We shake down acorns and pinenuts. We don't chop down the trees. We only use dead wood. But the White people plow up the ground, pull down the trees, kill everything. The tree says, 'Don't. I am sore. Don't hurt me.' But they chop it down and cut it up. The spirit of the land hates them. ... the White people destroy all. ... How can the spirit of the earth like the White man? ... Everywhere the White man has touched it, it is sore. (Wintu woman, quoted in McLuhan 1973: 15)

When people are exposed to market forces, in the face of inexorable globalisation, and are expected to measure and maximise profit financially, conflict may ensue. Intellectual property rights is one example, which increasingly concerns those of us working in the local knowledge in development field (where we currently have promotion of local science to complement global science), as development becomes evermore driven, like science, by market forces and business interests (see contributions by Clift and Bodeker, this volume). The sweeping up of people into the global market is giving rise to problems regarding 'ownership' of knowledge, challenging their commons approach to knowledge. The sharing of knowledge (which is also central to scientific practice) reflects a different value system to one that seeks to monopolise it. The market assumption is that we can buy knowledge and treat it as a commodity, patent it and count the profits. And if a company can exploit others' knowledge commercially without their knowledge, so much the better for the bottom line. An intriguing contemporary example concerns the *Hoodia* cactus of the Kalahari, slices of which the Khomani people eat to stave off hunger, and which a pharmaceutical company intends to develop into a drug to treat obesity. It is common knowledge among the Khomani. Initially left out of negotiations to patent the drug made from the cactus extract, they have lobbied for a royalty payment from the company involved that prospects to find and commercially exploit natural remedies (Wynberg 2004). Some may interpret this as evidence that they want a 'slice of the market action' but what such arrangements mean to them culturally is unclear and how they might use any income for the benefit of their entire community has yet to be worked out. It is concern to protect people's knowledge rights from pirate patents that has prompted India to launch a programme to create digital databases of local knowledge, although such databases can lead to problems of misrepresentation – even distortion – of knowledge by cutting it off from its bearers and 'freezing' it in time, whereas knowledge is dynamic and, part of culture, subject to constant refinement and change (Jayaraman 1999; Barr and Sillitoe 2000). But these attempts to record 'traditional knowledge' accord with unchanging, non-lineal worldviews.

The urge to count and measure has overflowed even further into areas where it is unhelpful in development contexts. We have agencies proposing measurable indicators of success that are difficult to comprehend, such as the Department for International Development's (DFID) expressed White Paper (2000: 12) target of reducing poverty by 50 percent by 2015 (following United Nations international development targets). We are unable to agree a definition of poverty. What is the difference between being poor and not poor? Is it plus one pence? We encounter another variant of Zeno's paradox. There is no such simple divide; poverty is a complex and relative condition influenced by a range of factors (which may make it an appealing target for measurement by politicians because they can fudge the results in any way they choose). The idea that we can measure poverty emanates from our belief that we should quantify wealth, a key aspect of our capitalist economic mindset, such that poverty is having a small amount of wealth, usually measured in money. Others teach us that wealth can be many other non-quantifiable things, such as strong family and kin networks, emotional security, spiritual well-being, freedom, non-stressful existence, and so on. 'Poor people' may be rich according to these criteria.

Some development programmes acknowledge that poverty is a complex condition, but the urge to find measurable 'outputs' to prove successful programme implementation is spawning dubious practices, such as the identification of social capital and cultural capital in the currently fashionable sustainable livelihoods approach, which it equates with financial capital and physical capital, and seeks to quantify (Carney 1998; Ashley and Carney 1999). It is a wholly bogus exercise. It is Einstein who is credited with observing that, 'not everything that can be counted counts, and not everything that counts can be counted' (*People and Plants Handbook* 2001: 26). The obsession with monitoring and measuring outputs and uptake to evaluate the success of development projects is inhibiting the practice of 'process' planning and implementation, where local people may meaningfully contribute to and influence the objectives and direction of projects. Instead 'blueprint' projects continue to dominate with top-down planning and control, often today disguised with participatory rhetoric. The belief in counting prevents local populations contributing significantly to their own 'development', when in reality it is not possible to measure empowerment, social changes, human improvement, and so on.

Statistics feature significantly, as in economics generally, in these efforts to measure development success and justify programmes politically. I have deliberately avoided such measures in this critique of the centrality of counting and measurement to Western understanding of the world, as too easy a target. We all know what we can do with statistics. It is possible to play all manner of games with them to demonstrate questionable propositions, such as cigarettes are good for one's health rather than bad for it (Brignell 2002). 'There are lies, damned lies and statistics', notes Moroney (1951: 2), who quotes the cynical adage in the opening to his classic introduction to statistics. For example, regarding the above-quoted statistic relating to DFID's poverty reduction goals, who calculated that 1.2 billion of people live in 'abject poverty' around the world, as it asserts? What is the worth of such a figure? It is at best a 'guesstimate'. And how on earth will DFID demonstrate that it has reduced this figure by 600 million (which is not to be cynical of the goal of reducing poverty, only the pretence at quantification)?

The need to acknowledge that knowledge and understanding, including that of science, are culturally conditioned and not value free is an increasingly critical issue with science, speaking a mathematical language that features infinity and zero, increasingly harnessed to the capitalist market with its unsustainable ends. We need to heed the implications. The capitalist ethos is a selfish one encouraging infinite demand. The ethos of New Guinea Highlanders is interesting here. They value sociopolitical exchange in opposition to market transaction. It allows for individual competition within communal contexts, structuring behaviour in sustainable ways to everyone's benefit (Sillitoe 1979). We find that wealth circulates around complex networks of kin relations, with no rich and poor poles, material winners and losers. Paradoxically, we might expect mania with measurement to have led our culture to a profoundly sustainable worldview, demonstrating the finite nature of resources, instead of the reverse. The reconfiguration of perspectives is not such a radical step. The other eco-discipline, the science of ecology, stresses the cyclic pattern of nature and informs demands for sustainable use of resources, energy generation and so on. It is the same if we contrast Christian values with capitalist values. According to the litany at the 2004 Founders and Benefactors commemoration service at Durham Cathedral, for instance, 'We look not to the things that are seen but to the things that are unseen; for the things that are seen are transient but the things that are unseen are eternal'. This ambiguous comment on the nature of the infinite and nothing catches nicely the tenor of our understanding of these polar concepts.

Notes

1. Quoted in Barrow (2000: 171–72).
2. As Schrader (1994: 510–11) notes, Newton 'stated specifically that his mathematical quantities were to be considered as described by continuous motion, not as existing in infinitely small parts. Leibniz, on the other hand, made the infinitely small quantities themselves the basic concepts in his differentials. Newton dealt with a finite quantity which is the ratio of two infinitely small quantities, the ratio of velocities; Leibniz dealt with the finite sum of an infinite number of infinitely small quantities'. Commenting on the calculus, Maor (1991: 18) observes that 'an infinite sum may add up to finite value, that is may converge to a limit … [terms] approach zero, but never actually become zero'.
3. As Horgan (1996: 203) notes, 'mathematics helps physicists define what is otherwise undefinable. A quark is a purely mathematical construct. It has no meaning apart from its mathematical definition. The properties of quarks – charm, colour, strangeness – are mathematical properties that have no analogue in the macroscopic world we inhabit'.
4. See Maor (1991: 235–36) for proof.
5. The largest prime number discovered to date is 2 to the 30, 402, 457th power, minus 1.
6. See Strathern (1977) for an account of calculations made by the Melpa in *moka* ceremonial exchanges, and Bowers and Lepi (1975: 312) among the neighbouring Kakoli; also Thune (1978: 74–78) for transactions among Normanby Islanders.
7. *Raskal* is the pidgin term for a criminal.
8. The Wola word for counting is *sa menay*.
9. Such four-base systems are common in the Southern Highlands, Western Highlands and Enga Provinces (Bowers and Lepi 1975: 315–17).
10. Nearby Kakoli speakers count in the same way (Bowers and Lepi 1975: 313).
11. While in keeping with the argument of this paper, this observation contradicts Mimika's (1988: 72, 122) suggestion that the Iqwaye can count to thousands and that their number system intimates familiarity with infinity. Mimika suggests that as the Iqwaye use the human body to count, each person's fingers and toes signifying twenty, the total number of people in the world, not only now but previously and in the future, represents infinity, for in 'this mythopoeic sense human population constitutes the set of "all possible human beings" – those that *are* and that yet *will be*' (1988: 74). He proceeds to draw parallels between such Iqwaye sets of humans (and sets of sets etc., arranged in twenty-base groups) with Cantor's set theory, which seeks formally to define infinity and manipulate it (Love 1994), in an exercise that confounds metaphor with correspondence, in maintaining that 'the Iqwaye system in its peculiar manner can be compared with the structure of infinite sets' (1988: 122).
12. I am grateful to Nigel Martin, Durham University Mathematics Department, for this example.
13. Such counting systems are fairly common, for example several Amazonian groups count in this way, including the Yanomamo, Siriona, Bacairi and Bororo people (Seife 2000: 7).
14. Vigesimal systems are common. For example in the Americas, the Maya (Seidenberg 1988), Inuit, Tlingit, Californian Indians, Navaho (Closs 1988) and Aztecs (Payne and Closs 1988) all had twenty-base counting systems, featuring the idea that one human being equalled twenty. Also in Africa, where the base-five building block is common, they occur among the Nuba of Sudan, Igbo and Yoruba of Nigeria, the Dyola of Guinea-Bissau, and the Banda of Central Africa (Zaslavsky 1973: 39–51).

15. Laycock (1975: 219–23) discusses others, and has a useful table comparing anatomical features with numbers in different languages.

16. The Melpa may likewise count large numbers in pairs, or eights or tens (Strathern 1977: 19); the Daribi routinely count in twos (Wagner 1967: 245).

17. For example 'little finger' is *jiy gongolow*, 'ring finger' is *jiy henday*, 'large finger' is *jiy ta*, 'index finger' is *jiy henday*, and 'thumb' is *jiy injiy* (*jiy* refers to 'hand'), which bear no relation to the numbers one to five: *mond, kab, teb, mak* and *suw* (see Bowers and Lepi 1975: 314, who report the same for the Kakoli).

18. Others have noted how the arrival of cash and associated decimal counting system have influenced, even undermined, traditional counting systems – Franklin and Franklin (1962: 191) among the Kewa; Bowers and Lepi (1975: 322) among the Kakoli; Hooley (1978: 157–58) among the Buang; Cheetham (1978: 23) among the Huli; Kettenis (1978: 32) among the Kilenge; and Thune (1978: 71) on Normanby Island.

19. The use of knotted devices occurs elsewhere too. In Africa, the Chagga of Tanzania and the Sundi of the Congo employed them (Zaslavsky 1973: 93–97), and they reach their zenith in the elaborate *quipus* knotted cords of the Inca (Ascher 1991: 15–26).

20. Similarly among the Imbonggu (Pumuge 1975: 21).

21. Likewise others in New Guinea, e.g. Kettenis (1978: 34) on the Kilenge, Smith (1978: 59) among the Kiwai, Thune (1978: 72) for Normanby Islanders, Vicedom and Tischner (1948: 239) for the Melpa, and Pumuge (1975: 21) for the Imbonggu.

22. Such numeral classifiers have been reported elsewhere in the world, and range from simple schemes such as that of the Wola, to complex ones, such as that of the Tzeltal Maya of Mexico who employ 528 numeral classifiers (Berlin 1968) and Dioi speakers of south China who have fifty-five of them (Ascher 1991: 12–14). The Aztecs had a similar system of numeral markers to the Wola which they used when counting different classes of things: *tetl* ('stone') for round things, *pantli* ('banner flag') for rows to things, *tlamantli* ('thing') for counting pairs, groups or different things, and *olotl* ('corn cob without kernals') for counting things that roll (Payne and Closs 1988).

23. The corollary is the possible denial of progress, a worldview that is the antithesis of development (Lee 1950: 95).

24. Today we even have computers set to calculate a final pi value, which seem arranged to calculate forever, spewing forth an incomprehensible post decimal point stream of numbers into infinity. In 2002 a team at Tokyo University announced that it had calculated the first trillion digits of pi, which took a supercomputer twenty-five days performing a trillion operations per second. The fundamental constants of science have a similar infinite quality, such as the fine structure constant a (7.29∞) the universal measure of the strength of nature's electromagnetic forces (Barrow 2000: 230). It is noteworthy that mathematicians have discovered that several infinite series have a relationship with pi, and that three-dimensional circular structures – the sphere and pseudosphere – have allowed them to overcome the Euclidean axiom that parallel lines will extend infinitely and never meet, and develop elliptic and hyperbolic geometry (Maor 1991: 34–39, 118–32).

References

Ascher, M. 1991. *Ethnomathematics: a Multicultural View of Mathematical Ideas.* Pacific Grove: Brooks/Cole Publishing Co.

Ashley, C. and D. Carney 1999. *Sustainable Livelihoods: Lessons from Early Experience.* London: Department for International Development.

Barr, J.J.F. and P. Sillitoe 2000. 'Databases, Indigenous Knowledge and Interdisciplinary Research', in *Indigenous Knowledge Development in Bangladesh: Present and Future*, (ed.) P. Sillitoe, pp. 179–95. London: Intermediate Technology Publications.

Barrow, J.D. 2000. *The Book of Nothing*. London: Jonathan Cape

Benardete, J.A. 1964. *Infinity: an Essay in Metaphysics*. Oxford: Oxford University Press.

Berlin, B. 1968. *Tzeltal Numeral Classifiers: a Study in Ethnographic Semantics*. The Hague: Mouton.

Biersack, A. 1982. 'The Logic of Misplaced Concreteness: Paiela Body Counting and the Nature of the Primitive Mind', *American Anthropologist* 84(4): 811–29.

Bowers, N and P. Lepi 1975. 'Kaugel Valley Systems of Reckoning', *Journal of the Polynesian Society* 84(3): 309–24.

Brignell, J. 2002. *Sorry, Wrong Number! The Abuse of Measurement*. Stockbridge, Hampshire: Brignell Associates (www.numberwatch.co.uk).

Carney, D. (ed.) 1998. *Sustainable Rural Livelihoods: What Contribution Can We Make?* London: Department for International Development (papers presented at International Development Advisers' Conference July 1998).

Cheetham, B. 1978. 'Counting and Number in Huli', in 'The Indigenous Mathematics Project', (ed.) D.F. Lancy. *Papua New Guinea Journal of Education* (special issue) 14: 16–27.

Closs, M.P. 1988. 'Native American Number Systems', in *Native American Mathematics*, (ed.) M. Closs, pp. 3–43. Austin: University of Texas Press.

Croal, P. and W. Darou 2002. 'Canadian First Nations' Experiences with International Development', in '*Participating In Development': Approaches to Indigenous Knowledge*, (eds), P. Sillitoe, A. Bicker and J. Pottier. London: Routledge (ASA Monograph Series No. 39).

DFID 2000. *Eliminating World Poverty: Making Globalisation Work for the Poor*. White Paper on International Development. London: HMSO.

Franklin, K. and J. Franklin 1962. 'The Kewa Counting System', *Journal of the Polynesian Society* 71: 188–91.

Gell, A. 1975. *Metamorphosis of the Cassowaries*. London: Athlone (LSE Monographs on Social Anthropology No 51).

Hilbert, D. 1964. 'On the Infinite', in *Philosophy of Mathematics*, (eds), P. Benacerraf and J. Putnam, pp.134–51. Oxford: Blackwell.

Hooley, B.A. 1978. 'Number and Time in Central Buang', *Kivung* 11(2): 152–70.

Horgan, J. 1996. *The End of Science*. Reading MA: Helix Books (Addison-Wesley).

Huxley, Aldous 1928. *Point Counter Point*. London: Grafton.

Janzen, J.M. and W. MacGaffey 1974. *An Anthology of Kongo Religion: Primary Texts from Lower Zaire*. Lawrence: University of Kansas Press (Publications in Anthropology No. 5).

Jayaraman, K.S. 1999. 'And India Protects Its Past on Line', *Nature* 401: 413–14.

Kettenis, F. 1978. 'Traditional Food Classification and Counting Systems of Kilenge – West New Britain', in The Indigenous Mathematics Project' (ed.) D.F. Lancy. *Papua New Guinea Journal of Education* (special issue) 14: 28–43.

Lancy, D.F. (ed.) 1978. The Indigenous Mathematics Project. *Papua New Guinea Journal of Education* (special issue) 14.

———— 1983. *Cross-cultural Studies in Cognition and Mathematics*. New York: Academic Press.

Laycock, D. 1975. 'Observations on Number Systems and Semantics', in *New Guinea Area Languages and Language Study*, (ed.) S.A. Wurm. Vol 1 Pacific Linguistics C38. Canberra: Australian National University [also at www.papuaweb.org/dlib/bk/pl/C38/_toc.html].

Lee, D. 1950. 'Lineal and Nonlineal Codifications of Reality', *Psychosomatic Medicine* 12: 89–97.

Love, W.P. 1994. 'Infinity: the Twilight Zone of Mathematics', in *From Five Fingers to Infinity*, (ed.) F. Swetz, pp. 658–67. Chicago: Open Court Publishing.

Maor, E. 1991. *To Infinity and Beyond: a Cultural History of the Infinite*. Princeton, NJ: Princeton University Press.

Mimika, J. 1988. *Intimations of Infinity: the Mythopoeia of the Iqwaye Counting System and Number*. Oxford: Berg.

Moore, A.W. 1990. *The Infinite*. London: Routledge.

Moroney, M.J. 1951. *Facts from Figures*. Harmondsworth: Penguin.

McLuhan, T.C. 1973. *Touch the Earth: a Self-portrait of Indian Existence*. London: Abacus.

Neihardt, J.G. 1961. *Black Elk Speaks*. Lincoln: University of Nebraska Press.

Panoff, M. 1970. 'Father Arithmetic: Numeration and Counting in New Britain', *Ethnology* 9: 358–65.

Payne, S.E. and M.P. Closs. 1988. 'A Survey of Aztec Numbers and Their Uses', in *Native American Mathematics*, (ed.) M.P. Closs, pp. 213–35. Austin: University of Texas Press.

People and Plants Handbook May 2001. Issue 6.

Price, D.J. and L. Pospisil 1966. 'A Survival of Babylonian Arithmetic in New Guinea', *Indian Journal of History of Science* 1(1): 30–33.

Pumuge, H.M. 1975. 'The Counting System of the Pekai-Alue Tribe of the Topopul Village in the Ialibu Sub-district in the Southern Highlands District, Papua New Guinea', *Science in New Guinea* 3(1): 19–25.

Rucker, R. 1995. *Infinity and the Mind: the Science and Philosophy of the Infinite.* Princeton, NJ: Princeton University Press.

Sabbagh, K. 2002. *Dr Riemann's Zeros.* Atlantic.

Salmon, W.C. (ed.) 2001. *Zeno's Paradoxes.* Indianapolis: Hackett Publishing.

du Sautoy, M. 2003. *The Music of the Primes.* Fourth Estate.

Saxe, G.B. 1981. 'Body Parts as Numerals: a Developmental Analysis of Numeration among the Oksapmin in Papua New Guinea', *Child Development* 52: 306–16.

Schrader, D.V. 1994. 'The Newton-Leibniz Controversy Concerning the Discovery of the Calculus', in *From Five Fingers to Infinity*, (ed.) F. Swetz, pp 509–20. Chicago, IL: Open Court Publishing.

Seidenberg, A. 1962. 'The Ritual Origin of Counting', *Archive for History of Exact Sciences* 2: 1–40.

———— 1988. 'The Zero in the Mayan Numerical Notation', in *Native American Mathematics*, (ed.) M. Closs, pp. 371–86. Austin: University of Texas Press.

Seife, C. 2000 Zero: *the Biography of a Dangerous Idea.* London: Souvenir Press.

Sillitoe, P. 1979. *Give and Take: Exchange in Wola Society.* Canberra: Australian National University Press.

Smith, G. 1978. 'Counting and Classification on Kiwai Island', in 'The indigenous mathematics project', (ed.) D.F. Lancy, *Papua New Guinea Journal of Education* (special issue) 14: 51–68.

Strathern, A. 1977. 'Mathematics in the Moka', *Papua New Guinea Journal of Education* 13(1): 16–20.

Swetz, F. (ed.) 1994. *From Five Fingers to Infinity.* Chicago, IL: Open Court Publishing.

Thompson, R.F. 1983. *Flash of the Spirit: African and Afro-American Art and Philosophy.* New York: Random House.

Thune, C. 1978. 'Numbers and Counting in Loboda: an Example of a Non-numerically Oriented Culture', in 'The Indigenous Mathematics Project', (ed.) D.F. Lancy. *Papua New Guinea Journal of Education* (special issue) 14: 69–80.

Vicedom, G.F. and H. Tischner 1948. *Die Mbowamb: die Kultur der Hagenberg-Stamme im Ostlichen Zentral Neuguinea.* Hamburg: Cram, De Gruyter and Co.

Wagner, R. 1967. *The Curse of Souw: Principles of Daribi Clan Definition and Alliance.* Chicago, IL: University of Chicago Press.

Williams, F.E. 1940. *Natives of Lake Kutubu.* Sydney: Oceania Monograph No. 6.

Wolfers, E.P. 1972. 'Counting and Numbers', in *Encyclopaedia of Papua New Guinea.* Melbourne: Melbourne University Press.

Wynberg, R. 2004. 'Rhetoric, Realism and Benefit Sharing: Use of Traditional Knowledge of Hoodia Species in the Development of an Appetite Suppressant', The Journal of *World Intellectual Property* 7(6): 851–76.

Zaslavsky, C. 1973. *Africa Counts: Number and Pattern in African Culture.* Boston, MA: Prindle, Weber and Schmidt Inc.

Index

Local science vs. global
science : approaches to
indigenous knowledge in
international development